PARADOXES OF PROGRESS

A Series of Books in Biology
EDITOR: Cedric I. Davern

PARADOXES

OF

PROGRESS

Gunther S. Stent
UNIVERSITY OF CALIFORNIA, BERKELEY

W. H. Freeman and Company
SAN FRANCISCO

Cover art on softcover edition reproduced from "The Golden Age," by Lucas Cranach. [Munich: Alte Pinakothek. By permission of Bayer. Staatsgemäldesammlungen.]

Library of Congress Cataloging in Publication Data

Stent, Gunther Siegmund, 1924–
    Paradoxes of progress.

    (A Series of books in biology)
    Includes bibliographies.
    CONTENTS: The rise and fall of Faustian man. The end of progress. The end of the arts and sciences. The road to Polynesia. Molecular genetics in the salon. What They are saying about Honest Jim. Prematurity and uniqueness in scientific discovery. Molecular biology and metaphysics. The dilemma of science and morals. [etc.]
    1. Science—Social aspects. 2. Science—Philosophy.
I. Title.
Q175.5.S74   301.24'3   78-17829

ISBN 0-7167-0080-8
ISBN 0-7167-0086-7 pbk.

Printed in the United States of America

2   3   4   5   6   7   8   9

Grateful acknowledgment is made to the publishers and journals listed below for their permission to reprint the selections included in this book:

Chapters 1 to 3: *The Coming of the Golden Age.* Copyright © 1969 by Gunther S. Stent. Reprinted by permission of Doubleday & Company, Inc. **Chapter 4:** "What They Are Saying about Honest Jim." *The Quarterly Review of Biology,* 43, 179–184, 1968. **Chapter 5:** "Prematurity and Uniqueness in Scientific Discovery." *Scientific American,* December 1972. Copyright © 1972 by Scientific American, Inc. All rights reserved. **Chapter 6:** "Molecular Biology and Metaphysics." *Nature* 248, 779–781, 1974, and "An Ode to Objectivity—Does God Play Dice?" *The Atlantic,* November 1971. Copyright © 1971 by The Atlantic Monthly Company, Boston, Mass. Reprinted with permission. **Chapter 7:** "The Dilemma of Science and Morals." *Genetics* 78, 41–51, 1974, and Zygon 10, 95–112, 1975. **Chapter 8:** Portions of this chapter are from "Cellular Communication." *Scientific American,* September 1972. Copyright © 1972 by Scientific American, Inc. All rights reserved. **Chapter 9:** "Explicit and Implicit Semantic Content of the Genetic Information." In *Logic, Methodology and Philosophy of Science.* R. Butts and J. Hintikka, eds. Dordrecht: D. Reidel, 1976. **Chapter 10:** "Limits to the Scientific Understanding of Man." *Science* 187, 1052–1957, 1975. Copyright © 1975 by the American Association for the Advancement of Science. **Chapter 11:** "The Poverty of Scientism." In *Foundation of Ethics and Its Relationship to Science.* H. Tristram Engelhardt and D. Callan, eds. Hastings-on-Hudson: Institute of Society, Ethics and the Life Sciences, 1977.

To Claire and Gabi,
Ronnie and Bob

Lass die Moleküle rasen,
Was sie auch zusammenknobeln!
Lass das Tüfteln, lass das Hobeln,
Heilig halte die Ekstasen!

[Never mind molecular gyration,
Whatever be its chance creation!
Stop perfecting, quit cerebration
Give ecstasies your veneration!]

CHRISTIAN MORGENSTERN
Galgenlieder, 1912

CONTENTS

PREFACE

This book is a collection of eleven essays on the social and philosophic impact of science—particularly the two life-science disciplines molecular biology and neurobiology, in which I have taught and carried out research during the past thirty years. In the first three essays I attempt to show why progress, that prominent Bicentennial feature of the (Western) human condition, to which science has contributed so heavily, is now nearing its end. In support of that apocalyptic conclusion, I draw attention to several internal contradictions—psychological, material, and cognitive—that render progress self-limiting. In the next four essays, I consider some insights that can be gained from the relatively brief history of molecular biology— still wholly within living memory; in fact, roughly contemporaneous with my own scientific career—into the problems posed by the sociology of science, the nature of scientific and artistic creativity, and the relation of science to ethics. Finally, in the last four essays I consider the dilemmas and paradoxes encountered in the attempt to give a scientific account of the nature of man. Here I draw attention to the technical and intellectual obstacles that lie in the way of understanding the brain and the processes that govern human ontogeny, from egg to adult. I also outline the fundamental epistemological limits that prevent us from coming to terms, from the viewpoint of science, with just those features of *Homo sapiens*, such as morally responsible personhood, that makes our species peculiarly human.

The first three essays were previously published as chapters of my earlier book *The Coming of the Golden Age,* and the remainder appeared originally as separate articles in various magazines and professional journals. In order to weave these essays into a more or less continuous narrative, I have somewhat altered and shortened their text.

I thank once more Jack Dunitz, Niels Jerne, Benoit Mandelbrot, and Ronald Stent, whose role as stimulating intellectual adversaries and constructive critics I previously acknowledged in the preface of *The Coming of the Golden Age.* In addition, I now express my indebtedness to David Hubel, Georges Kowalski, Allen Wheelis, and Torsten Wiesel, whose ideas and suggestions were crucial to the genesis of the more recent essays. Finally, I thank, as always, Margery Hoogs for her editorial assistance and her stylistically infallible red pencil.

March 1978   GUNTHER S. STENT

PARADOXES OF PROGRESS

"The Golden Age," by Lucas Cranach. [*Munich: Alte Pinakothek. By permission of Bayer. Staatsgemäldesammlungen.*]

PROLOGUE:

THE COMING
OF THE
GOLDEN AGE

The philosophical and sociopolitical discussions that in recent years have come to occupy more and more space in the pages of such erstwhile bastions of hard science as the journals *Science* and *Nature* presage that the history of post-Renaissance science—short if reckoned in sidereal time but immense if reckoned by the yardstick of secular change—is now reaching its ironical denouement. Contrary to expectation, it has turned out that the growth of our knowledge about nature has not made it any easier to reach rational decisions regarding man's fate. Instead, whereas the technological consequences of scientific progress have rendered the making of such decisions ever-more pressing and their effects ever-more grave, the intellectual consequences of scientific progress have made us aware of the difficulty, if not impossibility, of foreseeing the long-range results of our actions, while at the same time destroying the foundations for our judgment of their value. So what is to be done? Where do we go from here?

In his 1969 Presidential Address to the British Association for the Advancement of Science, entitled "On 'The Effecting of All Things Possible,'" Sir Peter Medawar, then director of the British National Institute for Medical Research, and supreme among contemporary scientist-philosophers for his combination of scientific excellence, erudition, and literary skill, took stock of this troublesome scene. The explicit purpose of Medawar's ad-

1

dress was "to draw certain parallels between the spiritual or philosophical condition of thoughtful people in the seventeenth century and in the contemporary world, and to ask why the great philosophical revival that brought comfort and a new kind of understanding to our predecessors has now apparently lost its power to reassure us and cheer us up." Medawar finds that both epochs—the first half of the seventeenth century and the second half of the twentieth—"are marked, not by any characteristic systems of belief . . . but by an equally characteristic system of unfixed beliefs; by the emptiness that is left when older doctrines have been found wanting and none has yet been found to take its place. . . . In the first half of the seventeenth century, the essentially medieval world picture of Elizabethan England had lost its power to satisfy and bring comfort, just as nowadays the radical materialism traditionally associated with Victorian times seems quite inadequate to remedy our complaints."

Then literature had, as it has now, an inward-looking character and a deep concern with personal salvation and establishment of personal authenticity. Then people were, as they are now, oppressed by a sense of decay and deterioration. Then there was, as there is now, a feeling of despondency and incompleteness, a sense of doubt about the adequacy of man. Then intelligent and learned men sought, as they do now, comfort and mystical syntheses between science and religion, because of the rootlessness or ambivalence of their philosophical thinking. But these cries of despair were not then, nor are they now, necessarily authentic; they tended, as they now again tend, to be an affectation of melancholy, a merely fashionable posture.

How did the English get out of their Jacobean rut? By having the first inklings of a new, hitherto unknown, cosmological concept: the idea of progress. This optimism arose, Medawar thought, from "the breathtaking thought that there was no apparent limit to human inventiveness and ingenuity. It was the notion of a perpetual *Plus Ultra*, that what was already known was only a tiny fraction of what remained to be discovered, so that there would always be more beyond."

So what went wrong? What is responsible for the sudden rise of the belief that progress has been a bust, that all that inventiveness, all that ingenuity, all those discoveries have brought us that monster, modern technology, which spawns new instruments of warfare and tramples down nature? How are we going to get out of our Space Age rut? By realizing, according to Medawar, that all this melancholy is quite uncalled for, being based on a superficial assessment of the human condition, and by hearkening, as did the Jacobeans, to the Trumpeter of Hope, Francis Bacon, who praised "the virtue and dignity of scientific learning and of its power to make the world a better place to live in." So take heart, men! After all, human history is only beginning; only during the past 500 years or so of their 500,000 year existence have human beings begun to be, in the biological sense, a success. Admittedly, Medawar "cannot point to a single definite solution of any one of the problems that confront us—political, economic, social, or moral . . . [But] we are still beginners, and for that reason may hope to improve. To deride the hope of progress is the ultimate fatuity, the last word in poverty of spirit and meanness of mind."

In the same year that Medawar delivered his homily, I brought out a counter-inspirational tract entitled, *The Coming of the Golden Age: A View of the End of Progress*, in which I drew conclusions diametrically opposite from Medawar's, and of which three chapters are reprinted in the first part of this collection of essays. Rather than viewing the abandonment of the hope of progress as the ultimate fatuity and declaring that what was good enough for the Jacobeans is good enough for us, I concluded that progress really *is* nearing its end in our time. Thus the waning belief in progress would reflect an accurate assessment of the actual situation, rather than a poverty of spirit and meanness of mind. My principal reason for this conclusion was that progress embodies several internal contradictions—psychological, material, and epistemological—which render it self-limiting. Thus, in contrast to Medawar's stated belief that there is no "intrinsic limitation upon our ability to answer questions that belong to the domain of natural knowl-

edge and fall, therefore, within the agenda of scientific inquiry,"
I maintained that there *do* exist some fundamental abrogations
of that agenda. As I tried to show, much of the knowledge and
know-how which we still lack in the management of human
affairs has so far eluded our ken, not because we are still beginners with hope for improvement, but because it pertains to
phenomena whose causal connections posited by our imagination are not susceptible of validation.

The main thesis of my book was that we are about to enter a
Golden Age, upon the coming of which the arts and sciences
will have reached the end of their long road. The Golden Age
to which I referred is that of Greek mythology, recorded by
Hesiod in the eighth century B.C. According to this myth, the
present obviously miserable Iron Age is but the fifth stage in a
constantly deteriorating series of stages, the first stage of which
was the Golden Age. In that Golden Age, a golden race of
mortal men dwelt on Earth, who "lived like gods without sorrow of heart, remote and free from toil and grief, miserable age
rested not on them, but with legs and arms never failing they
made merry feasting beyond the reach of all evil. When they
died, it was as though they were overcome with sleep, and
they had all good things; for the fruitful earth bore them fruit
abundantly without stint. They dwelt in ease and peace upon
their lands with many good things, rich in flocks and loved by
the blessed gods." This Golden Age, according to Hesiod,
presently came to an end when Pandora lifted the lid of her
box and allowed the escape and spread of previously unknown
evils. The Golden Age was then succeeded by the Silver, Brass,
and Heroic Ages, each age worse than its predecessor, and finally by our own Iron Age. In our own Iron Age men "never
rest from labor and sorrow by day, and from perishing by
night; and the gods shall lay sore trouble upon them."

The purpose of my essay was to show that the ancients' view
of human history was topsy-turvy, in that the Golden Age is
not the very first but the very last stage of history, and one that
is a necessary successor, rather than an antecedent, of the Iron
Age. I tried to show that unmistakable signs of the advent of
the Golden Age, and all that it portends, are already with us,
at least in the industrially advanced nations. I doubted, how-

ever, that Hesiod, or any other of the legion of writers who pined for lost paradises since his time, would find this Golden Age very much to his liking.

My general argument followed more or less Hegelian (or, for all I know, Marxist) lines. I tried to show that internal contradictions—theses and antitheses—in progress, art, science, and other phenomena relevant to the human condition make these processes self-limiting; that these processes are reaching their limits in our time and that they all lead to one final, grand synthesis, the Golden Age. On first sight, the finding that progress in general and creative activity in particular are now reaching their ends might have appeared to reflect a deeply pessimistic outlook on the future, a typical product of Nuclear Age *Weltschmerz*. On second sight, however, it should have become apparent that my conclusions were, if anything, optimistic, since I showed that just at that very moment in history when the possibilities for future progress and creative exertion are becoming exhausted, the secular consequences of past progress have given rise to a human psyche which is perfectly adapted to that entirely novel condition. This view is, therefore, in harmony with the precepts of Voltaire's Doctor Pangloss, since where else could such amazingly felicitous concord have occurred but in the best of possible worlds?

The Coming of The Golden Age did not receive wide critical notice. It was reviewed only by a few friends and colleagues, such as Jerome Lettvin, Rollin Hotchkiss, and Max Delbrück, praised in an anonymous paragraph in *The New Yorker*, and panned by an M.I.T. student in a sophomoric piece in *The Nation*. In view of this general lack of attention, I was pleased to discover that in its issue of November 3, 1971, the Moscow *Literaturnaya Gazeta*, with the largest circulation of any literary weekly in the world, published a discussion of my book by V. Kelle, a member of the Institute of Philosophy of the USSR Academy of Sciences, under the Baconian title, "The Road to Knowledge Has No End." Kelle wrote that "according to G. Stent, science is now reaching its end because the 'intellectual gains' which it provides are decreasing; so society loses its interest and stops supporting science financially. It is very characteristic that G. Stent connects this coming end of scientific progress with the transformation of society into one in which the interest in science will no longer be completely determined by business interests. Thus, G. Stent admits that capitalism considers science only from a utilitarian viewpoint and regards it merely as a sphere for profitable investment." Although, according to Kelle, "not only G. Stent but many other scientists in the West realize the straightjacket into which the capitalist economy for profit has squeezed the modern scientific revolution, it is a pity that not many of them can draw the right socio-political conclusions from this insight." For instance, G. Stent makes the error of inferring the existence of intrinsic bounds to the growth of knowledge due to an eventual exhaustion of the subject and the limitations of the human mind, because he does not understand that according to Marxist philosophy learning moves not in circles but in spirals. If G. Stent were not so innocent of the materialist-dialectical approach, he would know that "the very contradiction between the bounds of knowledge at every stage of historical development and the unlimited possibility of knowledge in general is what moves science. This contradiction, which, in the end, always results in a scientific advance, arises again and again. Though sciences may come and go, science is eternal."

In any case, Kelle finds my discussion of the end of progress more interesting from the psychological than from the philosophical point of view. "What is important here is the mood of the author himself. What caused such a mood, such a way of perception?" The answer is obvious: "the conditions of life, the sentiments of the society in which he lives." How did these sentiments arise? "After the War, a pessimistic mood characterized American society. Eventually, however, that bourgeois society understood that it cannot develop without ideals and that if it does not provide these ideals it will be difficult to hold the young generation within the necessary social frame." So, according to Kelle, to keep the young in line, the ideal of the post-industrial society was cooked up, which tries to make believe that the United States and other developed capitalist countries have reached the top rung of the staircase to social progress. This thoroughly bourgeois concept "assigned a ruling position to science and to the intellectual elite connected with it.

According to this view, what moves mankind forward, settles troubling problems and makes society healthy is not the revolutionary movement of the masses but the fulfillment of technical decisions of specialists acting within a framework of rational organization." Naturally, as could have been foreseen by any Marxist philosopher, this false ideal of progress via the so-called post-industrial society could not satisfy the needs of bourgeois society. Why? Because "the bourgeoisie turns out to be scared by the rapidity of modern socio-economic development and is lost in the face of the perspectives of the future. It sees the wheel of history spinning faster and faster but cannot see where it is spinning to. To what will all that progress lead? Isn't it approaching some critical point, at which everything will crash? It is this fear for the future that has brought forth wishful thinking: would it not be better if the development of science, and hence of the economy, began to decline gradually? That is the social soil on which ideas like those of G. Stent grow."

In contrast to Kelle's perception of my views of the intrinsic limitations of scientific progress as wishful thinking, and hence as *optimistic* within the context of the future-fearing social soil on which I live, in the only other published discussion of my book, a fellow-bourgeois biologist, Bentley Glass, perceived them as "pessimistic visions of man's future." In his 1971 Presidential Address to the American Association for the Advancement of Science, "Science: Endless Horizons or Golden Age?" Glass considers the possibility of limits to scientific understanding. Glass' titular question is a composite of the titles of Vannevar Bush's *Endless Horizons* and my book, chosen as representatives of opposite extremes "in the spectrum of belief in the future of science—the one, the view of limitless expanding knowledge and of infinite bounds, the other, the view that scientific knowledge, like our universe, must be finite and that the most significant laws of nature will soon have been discovered." What would Kelle have said about *Endless Horizons,* written in the immediate post-War period by an American Elder Statesman of Science? Bush believes, as does Kelle, that no end of the road to knowledge is in sight. But Bush claimed that it is an intellectual elite of men of rare vision—the "builders"—rather than the movement of the masses that will convert an indefinitely long succession of scientific advances into social progress. Aha! Wisened by Kelle to such bourgeois misconceptions we have no trouble identifying Bush as one of the creators of the myth of the post-industrial society invented during the pessimistic post-War years to keep American youth in line. So from the dialectical-materialist point of view, *Endless Horizons* and *The Coming of the Golden Age* are only superficially in diametric opposition: deep under, both represent the same kind of bourgeois wishful thinking, propagated to ward off the inevitable collapse of capitalist society.

In fact, Glass also perceives that, *au fond*, Bush's ideas about the limits to scientific progress and mine are not all that different. As Glass shows, Bush's notion of science was not really that of an infinitely large reservoir of discoverable knowledge, but rather that of a structured edifice of finite extent. That being the case, Glass asks: was not Bush's metaphor of the Endless Horizons "supposed merely to imply that from our present viewpoint so much yet remains before us to be discovered that the horizons seem virtually endless?" But then "with each new phenomenon discovered and explored, with each new law confirmed, there is an approach to the finite limits of scientific knowledge. In that case it is less important to note the absolute bounds of knowledge at the present time than to examine the rate at which, in the past century and a half, our scientific knowledge has been expanding." That this rate has been increasing exponentially, and hence that the approach to the limits has been ever accelerating, is one of the central claims of the *Coming of the Golden Age*.

I think it can fairly be said that since the late 1960's the Baconian faith in progress through science with its endless horizons has undergone a further decline. For instance, the wide appeal and popularity of such more recent anti-progress writings as Meadows' *The Limits to Growth: A Report for the Club of Rome's Project on the Predicament of Mankind* and Roszak's *The Making of the Counter Culture: Reflections of the Technocratic Society and Its Youthful Opposition* and *Where the Wasteland Ends: Politics and Transcendence in Postindustrial Society*, reflect this change in spiritual climate, as do the previously all-but-inconceivable demands by important sectors of the bourgeois society for the abandonment of highway, bridge, dam, power plant and supersonic aircraft construction projects. (Kelle's dialectics notwithstanding, such anti-progressive sentiments now appear to be growing also on the social soil of the USSR, and thus can evidently arise just as well via the revolutionary movement of the masses.)

The recent fate of my own professional specialty, molecular genetics, to whose rise and fall I devoted a large part of *The Coming of the Golden Age*, has provided one of the more dramatic indicators of this growing loss of the will for "The Effecting of All Things Possible." The great advances that had been made by the late 1960s in understanding hereditary processes now held out the promise for the development of a technology of "genetic engineering" that would be bound to produce tremendous benefits for agriculture, medicine, and other domains affecting human welfare. Accordingly, Bentley Glass closed his 1971 Presidential Address by expressing his hope for a future improvement of the physiological and psychological characteristics of mankind via a judicious modification of the human hereditary material. But in 1975, just when the direct chemical manipulation of hereditary material had become a practical possibility, the scientists mainly responsible for this development held a widely-publicized international conference to draw attention to the dangers inherent in their work. Thus there came about the paradoxical situation, well-nigh unprecedented in the annals of modern science, that the very persons who labored so hard to advance scientific research suddenly began to view themselves as potential sorcerer's apprentices, unable to control the agents they created, at the very moment when tremendous results seemed finally within their reach. Almost overnight, the public image of genetic engineering was altered from that of a great potential benefaction to mankind to that of a sinister project of mad scientists willing to risk the welfare of humanity for the satisfaction of their own idle curiosity. In 1976, I imagine, Bentley Glass would have thought twice before putting forward the kind of genetic recommendations for man's future that seemed perfectly reasonable for him to make only five years earlier.

Thus feeling that the subsequent course of events confirmed the essential validity of the prognoses I made in *The Coming of the Golden Age*, I tried to develop further my earlier ideas on the social and philosophic aspects of science and its future. Although I believe that my basic outlook on these matters has undergone little change since the late 1960's, my focus of attention has shifted somewhat, for several reasons. First, I abandoned molecular genetics, my earlier field of specialization, and like other veterans of the Romantic Period of molecular genetics, took my orders as a neurobiologist. Second, I became aware of the importance of the structuralist approach to man in such disciplines as psychology, anthropology and linguistics. Third, I came to realize that the epistemological paradoxes which I had earlier seen to constitute limitations for scientific progress have their parallel in the ethical domain and that the tremendous recent increase in man's mastery over nature has brought these hitherto mainly hidden contradictions into the open. Finally, I became more familiar with the history of Chinese philosophy and science and realized that the "Wisdom of the East" is nothing other than an attempt to resolve these contradictions by restructuring the ensemble of human aims. Accordingly, the themes of neurobiology and structuralism, epistemological and ethical paradoxes, and Taoism and Confucianism recur, fugue-like, throughout the essays that make up the remainder of this book.

Alas, this narrative has no close; it does not end with an inspirational message, does not put forward a eudaemonic course of action for the salvation of mankind. Like *The Coming of the Golden Age*, it merely reiterates the insight passed on to us from the author of the *Book of Genesis*, via Søren Kierkegaard and Friedrich Nietzsche, but repressed by the utopian visionaries from Thomas More and Bacon to Marx, that man's troublesome existence arises from his paradoxical nature: half bestial, half divine. *Homo sapiens* may return to Eden, but not man.

Bibliography

Bari, G. "An End to Progress." *The Nation*, October 19, 1970, p. 380.

Bush, V. *Endless Horizons*. Washington D.C.: Public Affairs Press, 1946.

Delbrück, M., and R. E. Dickerson. "A Double Review." *Engineering and Science* 33, 53–55 (1970).

Glass, B. "Science: Endless Horizons or Golden Age?" *Science* 171, 23–29 (1971).

Hesiod. *The Works and Days.*

Hotchkiss, R. "Almost There." *Science* 169, 664–666 (1970).

Lettvin, J. "The Rise and Fall of Progress." *Natural History,* March 1970, pp. 80–82.

Meadows, Donella H., D. L. Meadows, J. Randers, and W. W. Behrens, III. *The Limits to Growth* (A Report for the Club of Rome's Project on the Predicament of Mankind). New York: Universe Books, 1972.

Medawar, Sir Peter. "On 'The Effecting of All Things Possible.'" *The Advancement of Science* 26, 1–9 (1969–1970).

The New Yorker, April 18, 1970, p. 168.

Roszak, T. *The Making of a Counter Culture: Reflections on the Technocratic Society and Its Youthful Opposition.* Garden City: Doubleday, 1969.

Roszak, T. *Where the Wasteland Ends: Politics and Transcendence in Postindustrial Society.* Garden City: Doubleday, 1969.

Stent, G. S. *The Coming of the Golden Age: A View of the End of Progress.* Garden City: The Natural History Press, 1969.

I

THE RISE
AND FALL
OF
FAUSTIAN MAN

"Dr. Faustus in His Study," by Rembrandt.
[*By permission of The Bettman Archive, Inc.*]

1

THE END
OF
PROGRESS

[*1969*]

In the early 1950s the beatniks suddenly made their appearance in the North Beach district of San Francisco. On first sight this phenomenon seemed to represent a revolt against the contemporary standards of middle-class America. Slovenly, sandal-shod young men and women gathered on Upper Grant Avenue to lead what appeared to be a dissolute life, in flagrant negation of the very values of the wholesome environment whence these youths had sprung. The beards and long hair of the men made them stand out from the clean-shaven, crew-cut All-American boys, though a century earlier their tonsorial habits would have made the beatniks blend in perfectly with the North Beach scene of the pioneer forty-niners. And the abnegation of lipstick and rouge set off the women from the cosmetic radiance of the All-American girls. The public attitude toward the beatniks was either an illiberal, uncomprehending hostility, or a liberal, amused tolerance, based on the understanding that revolt against parental authority and custom is a natural, and perhaps even a healthy thing. In any case, most people believed that as soon as these aberrant youths reached middle age, they would mend their ways and buckle down to the job of getting on in life. In apparent confirmation of this prognosis, the beatniks *had* disappeared from North Beach by the 1960s, at which time their former haunts came to be occupied by restaurants and gift shops catering to tourists and other solid citizens. The beatniks seemed, therefore, to have been just one more variety of Bohemians, who seem to come and go as bizarre fringe phenomena of the mainstream of social and cultural evolution.

But as its history shows, it is a mistake to dismiss Bohemianism all that lightly. In retrospect, the Bohemians of Europe and America seem to have been recruited from among the most sensitive and brightest youth of their generation. The Bohemians saw sooner and more clearly than their less perceptive contemporaries the contradictions of their surroundings and adopted radical solutions to the paradoxes of facts and mores which faced them at the threshold of their adulthood. Contrary to the simplistic view, Bohemians have generally not abandoned at all their radical attitudes and tastes upon reaching middle age and eventual reintegration into society. Instead, it usually was society which meanwhile had changed and come to assimilate what were once far-out notions. Seen from this viewpoint, Bohemians represent a vanguard whose present radical mores can serve as a prospectus for the future conventional mores of the Establishment. For instance, a retrospective look at the very first Bohemians of nineteenth-century Montmartre shows that their artistic tastes and standards of personal behavior—so *épatant* to their contemporary bourgeois—presently became accepted middle-class values of post-World War I Europe. Another example is that of the American post-World War I Bohemia of Greenwich Village. Here were then gathered young people who were repelled by the dog-eat-dog social Darwinism of American capitalism and by the venal vulgarity of its aesthetic standards. The Greenwich Village left-of-center politics and rejection of the almighty dollar as the alpha and omega of goodness were to become accepted values of the American post-World War II Establishment. The veterans of Greenwich Village did not need to conform to society upon reaching middle age; by then, society had already conformed to *their* standards. And so it would have been worthwhile to examine the philosophy of the beatniks in the 1950s if one had wanted to get a preview of what metropolitan America of the 1960s would be like.

Beat philosophy represents a rather radical departure from post-Renaissance Western attitudes, though it seems conventional from the purview of Oriental thought. It renounces both use of reason and striving for worldly success, which are felt to be irrelevant for, or even obstacles to, true living. That is, beat philosophy asserts that feelings and immediate sense experiences should take precedence over cerebration, and that realization of the self is to be sought in inner-directed rather than outer-directed exertions. By the 1960s, the beatniks had faded from view, not because they had actually disappeared but because their attitudes and styles had become a commonplace in the metropolitan areas of the East and West coasts. No one turned around any longer to take a second look at a beard or a sandal. Meanwhile, beat philosophy had moved across San Francisco Bay and matriculated in the University of California at Berkeley, though this fact had not been noticed by its then administration. It took the trauma of the Free Speech Movement to call attention to the profound change that had come about in the nature of the Berkeley student body. As the Muscatine Report of the University's Academic Senate found, an ever-increasing number of the better students no longer appeared to be "academically oriented," or "fixed upon careers and . . . seizing the opportunities offered by the University to educate themselves for a life-time of work and advancement in their fields." A group of "nonconformist" students had come to the fore, whose "most obvious feature of their outlook . . . is their outright rejection of many aspects of present-day America." These students believed that Americans "who claim to be moral are really immoral, and those who claim to be sane are truly insane . . . These ways of rejecting society in one's private life are outgrowths of the patterns of the earlier beat, or noncommitted generation." But the *really* radical aspect of the new student mentality was not the superficially obvious and by no means novel attitude of social protest but its underlying antirational basis. For "students who hold the belief that feeling is a surer guide to truth than is reason cannot readily appreciate the University's commitment to rational investigation."

I now intend to inquire into the origins of those twin aspects central to beat philosophy, the antirational and the antisuccess, which had become manifest among Berkeley students. It should be made clear at the outset that the antisuccess aspect goes far beyond opposition to meretricious strife for material reward—by the late 1950s *that* kind of success was already deflated even in nonbeat, or square society—but extends to any and all achievements in the outer world. Thus the writers of beat philosophy who set forth these notions—Kerouac, Ginsberg, Mailer, for instance—could not have been wholly beat themselves because a beat littérateur is a contradiction in terms. This same contradiction by the way, is encountered in Zen Buddhism, to which beat philosophy is strongly affined. For it is claimed by Zen masters that no person can have *really* understood Zen if he so much as attempts to write about it.

The Muscatine Report makes what I believe to be the correct identification of the source of the rise of the beat attitude among the post-World War II generation: the Affluent Society. Growing up in a society from whose ethos poverty and want have been banished, and in which a basic level of economic security is taken for granted, engenders a beat psyche to which strife for success is largely foreign. For success is a goal imbibed in a childhood spent in an *ambiance* of paramount economic want and insecurity.

I will now try to justify this portentous inference in terms of the rather old-fashioned, nineteenth-century concept of the *will to power.* This concept was central to the philosophy of Nietzsche, who considered it as the metaphysical essence of life itself. According to him, wherever there is life, there is will to power. In order to avoid such metaphysical notions, however, I shall treat the will to power simply as psychological fact; namely, I will take it for granted that in the human psyche there exists a will to have power over the events of the outer world. And following Nietzsche, I shall adopt the view that sublimation of that will to power is the psychological mainspring of all creative activity. Undoubtedly, the will to power concept can be restated more satisfactorily in modern psychoanalytic terms, as a dynamic relation between ego and id. But I think for our present purposes it is not necessary to probe into the relative importance of conscious and subconscious components of that will. In any case, the will to power is patently one of the most important driving forces behind our outer-directed behavior. We may now define success in the broadest sense as an exercise of the will to power in which the self finds that the results which it expected from that exercise were actually met. That is, success means the ability to manipulate the events of the outside world in a satisfactory manner. These subjective findings in relation to the exercise of the will to power exert an important feedback on the self, which finds realization in terms of the success of its outer-directed behavior.

We may now inquire into the biological origin, both ontogenetic and phylogenetic, of the will to power—an inquiry that is not, of course, meaningful from Nietzsche's metaphysical standpoint. For this purpose we may envisage that the will to power has both innate, or instinctive, and received, or learned components. Its ontogenetic origin is, therefore, to be sought in an interaction between genetically determined concepts inherent in the structure of the human brain and experiential notions acquired after birth. That is to say, the will to have power over the events of the outer world does not arise automatically in infancy; instead the intensity and particular form of its development depends on ideas received from the childhood environment. And as far as its phylogenetic origin is concerned, it would follow that the will to power—a peculiarly human attribute whose appearance must have been a crucial step in the hominization process—arose through the natural selection of behavior. That is to say, according to the "survival of the fittest" principle, natural selection favored those proto-human genes which produce a brain in which the will to power concept is innately latent. Concomitantly with this process there occurred also a selection of those proto-human groups which propagated the ideas necessary for converting the latent into the overt will to power. This argument bears some considerable affinity to Noam Chomsky's theory of the origin of linguistic capacity. For Chomsky proposes that the structure of the human brain embodies within it a "universal" grammar on the basis of which the "particular" grammars of all natural languages have been generated and thanks to the instinctive knowledge of which the child is able to master the otherwise well-nigh impossible feat of recognizing the logical structure of the word sequences spoken by the adults of his environment. From this point of view, the acquisition of language is the product of an interaction between received particular ideas and an innate general logical system. This system is attuned to these ideas precisely because they were generated by a homologous system in the first place.

Here we have touched on a special feature of human evolution. Since man passes on to his offspring not only genetic traits but also ideas, natural selection operates on him also at the paragenetic, ideational level and favors the survival of those groups which propagate the fittest ideas. [It is to be noted in this connection that the stability of the evolutionary transmission of an idea is the greater the earlier it can be transmitted in child rearing and the higher its affective content.] Thus, the early transmission to the child of the will to power of its parents is bound to have had great adaptive survival value in a generally hostile environment. The stability of this transmission is assured also by incorporating the will to power idea into the social ethos which the child imbibes, for instance, through fables such as La Fontaine's "The Grasshopper and the Ant." In having fostered the cultivation of such attributes as curiosity, ambition, and imagination, the will to power provided man with the psychological wherewithal to gain ascendancy over his fellow creatures.

Indeed, the very roots of rational thought are likely to lie in the will to have power over the events in the outer world. For the notion of interpreting the events of that world in terms of postulated causal connections must have been one of the most highly adaptive ideas in the whole of human evolution. Here, however, we may already note one self-limiting product of rational thought which can exert a negative feedback on the will to power: the idea that God's will forms part of these causal connections. That is, the greater Divine intervention in the events of the outside world and the smaller the influence of mortal will on Divine will, the less scope there is for any exercise of the will to power. I think it is possible that the gradual hegemony of this fatalistic intellectual short-circuit may have been responsible for the weakening of the will to power that appears to have taken place in such earlier theocratic civilizations as Egypt and Byzantium. That the manner of realization of basic human drives *is*, in fact, subject to evolution, and that the alleged constancy of "human nature" is a fallacy, was recognized by Nietzsche and by Herbert Spencer more than a hundred years ago. Since that time the results of comparative ethnology have amply demonstrated human adaptive processes at the paragenetic, ideational level. And, to continue belaboring the obvious, we might note that the speed with which ideational adaptation can respond to a change in selective conditions is very much greater than the possible rate of genetic adaptation. For ideas can be gained or lost and can spread through populations much more readily than can the DNA nucleotide base sequences of genes.

With the rise of civilization in the Fertile Crescent ten thousand years ago, there became possible a sublimation of the exercise of the will to power into higher spheres of creative activity, whose exclusive concern no longer had to be the problem of the next meal. In exercising the sublimated will to power, the manipulation of external events has become an end in itself. Here the self no longer finds success on the basis of mere gratification of physiological needs but evaluates whether it has managed to bring about the intended change in the outside world. Finally, this sublimation culminated in the archetype whom Oswald Spengler called *Faustian Man*. The boundless will to power of Faustian Man causes him to view himself as being locked in an endless strife with his world to overcome obstacles, conflict, to his mind, being the very essence of existence. Thus Nietzsche's metaphysic of the will to power is the philosophy of Faustian Man. Since Faustian Man reaches for the infinite, he is *never* satisfied. His personality is endowed with lifelong growth, since, never finding satisfaction in success, the Faustian self ceaselessly seeks further realization through outer-directed activities. In my further exposition, I shall adopt Oswald Spengler's characterization of Faustian Man, the epitome of the will to power, as the prime creative mover of history.

Different individuals in the same society are obviously possessed of the will to power to different extents, differences which one may attribute in large part to variations in the manner and environment of child rearing. (However, congenital, physiological variations undoubtedly also play a role in these differences.) Concerning these differences, Ortega y Gasset has

said that "the most radical division that it is possible to make of
humanity is that which splits it into two classes of creatures:
those who make great demands of themselves, piling up difficulties and duties and those who demand nothing special of
themselves, but for whom to live is to be every moment what
they already are, without imposing on themselves any effort
towards perfection, mere buoys that float on the waves." As
individuals can differ in the intensity of their will to power, so
can societies differ in the intensity and manner of distribution of
that will to power among its members. And here it seems to me
that a most important factor influencing that intensity is the
degree to which the awareness of economic insecurity forms
part of the ethos of each society. I do not yet know how to justify
fully this probably Marxist view except by making the obvious
point that the higher the degree of economic insecurity extant,
the greater the power over external events needed by the individual for his survival. Thus the realization of this commonplace
by parents and educators provides strong incentive to make the
transmission of the will to power an important factor in child
rearing. This attitude finds reinforcement in early childhood in
exposure to the success-oriented homiletic content of epics,
folklore, and fairy tales and in adolescence by encounter with
the situation of the real world. It is to be noted, however, that
from these considerations it does not automatically follow that
in a society of general want it is the children of the poor who
develop the most and the children of the rich who develop the
least will to power. First, children of both rich and poor imbibe
the ethos of their society and, second, some rich become, or
remain rich, because of the greater than average intensity of
their will to power and thus expose their own children to this
quality. On the other hand, the familiar dissolute wastrel offspring of the rich can be thought to *have* arisen from a break
in transmission of the will to power. In any case, the *adaptive
value* of the will to power would be strongly diminished in an
environment from which economic insecurity had largely
disappeared.

One case of an earlier civilization in which a lessening of the
will to power can be thought to have been engendered by
economic causes, and of particular interest to us here because

of its affinity to the beat scene, is that of the flowering of Zen Buddhism in seventh-century T'ang China. During the T'ang dynasty, China came to know a degree of internal security and economic stability previously unattained in the history of mankind. I think the emergence under these circumstances of an anti-Faustian philosophy which emphasizes the realization of self through inward-directed processes rather than through power over the outside world offers good support to the notion that economic insecurity is an important factor for the maintenance of the will to power, and hence for the perpetuation of Faustian Man. In Western society, a decline of Faustian Man set in the nineteenth century, mainly brought about by the economic fruits of the Industrial Revolution and the social consequences of the rise of liberal democracies in Europe and America. The ever-mounting degree of security provided to the citizens of bourgeois societies then began a gradual erosion of the intensity with which the environment of child rearing engendered the will to power in the adult.

One of the first signs of this gradual change in the motivational makeup of Western man was the decline of romanticism. Early in the nineteenth century romanticism had celebrated the final apotheosis of Faustian Man in the works of Schopenhauer, Goethe, and Beethoven. By the latter part of the nineteenth century, the romantic notion of the self as a free agent exerting will to power on obstacles of the outer world was giving way to inner-directed searches for the self, as reflected in the works of Kierkegaard and Dostoevski. Concomitantly, the Faustian ideal of the rugged individualism of laissez-faire capitalism was slowly making way for the anti-Faustian gospel of socialism, under which the individual gains his identity mainly from class membership and has little freedom other than playing out his role in the dialectics of class struggle. But these anti-Faustian philosophical and political notions of an intellectual avant-garde were only the first swallows of a new spring. In any case, their authors must have still retained a large Faustian element in their own makeup to have been able to create their works. After World War I, the Faustian decline had become more evident in Europe. Now talk of "decadence" and of the "decline of the West" became part of the Zeitgeist of the interwar European intelligentsia. This general climate provided fertile ground for Oswald Spengler's and Arnold Toynbee's views of history in terms of an inexorable rise and fall of civilizations. But the return to economic insecurity brought by the Great Depression and the throwback into barbarism brought by the rise of fascism—thus the *de facto* advent of a retrogression of civilized life—temporarily blew away such sentiments of European *Weltschmerz*. In America, however, where despite its higher standard of living the feeling of general economic security was much less in evidence than in Europe, there had been no analogous widespread feeling of impending decline after World War I. Only with the end of the Great Depression in the late 1930s did there begin in America that period of continuous prosperity which culminated in the postwar Affluent Society. The New Deal, World War II, and advances in technology raised the general standard of living to previously unknown heights, and the specter of the struggle for economic survival had vanished. Economic well-being was now taken for granted. This change in the economic situation engendered a corresponding change in social ethos which deflated the idea of success. And this change in ethos, it seems patent to me, caused a massive reduction of the will to power in the first generation to be reared under its ambiance: the beat generation. A few years later, affluence following postwar reconstruction seems to have set off an analogous motivational metamorphosis also among European youth, in both capitalist West and Socialist East.

And here we can perceive an internal contradiction of progress. Progress depends on the exertions of Faustian Man, whose motivational mainspring is the idea of the will to power. But when progress has proceeded far enough to provide an ambiance of economic security for Everyman, the resulting social ethos works against the transmission of the will to power during child rearing and hence aborts the development of Faustian Man. This internal contradiction thus embodies in progress an element of negative feedback. A formally analogous analysis of such an internal contradiction in progress was made by Ortega y Gasset. He recognized Faustian Man as being the mainspring of progress and believed that economic security leads to the hegemony of a non-Faustian mass man. Ortega y Gasset developed the idea that the culmination of the fantastically successful efforts of Faustian Man in the eighteenth and nineteenth centuries allowed the non-Faustian, noncreative masses to take over in the twentieth. That is, the combination of economic prosperity and of liberal notions, such as the rights of man, promulgated by earlier Faustian leaders finally gave power to the non-Faustian masses. The mobocracy of the masses, who had previously accepted their inferiority in docile obscurity, now smothered their benefactor, Faustian Man. Though I do not, of course, adopt the aristocratic frame of reference of Ortega y Gasset's argument, its net result is essentially the same as that which I am trying to make here.

Thus I reach my first general conclusion concerning progress: It is by its very nature, by its very dependence on the will to power, *self-limiting*. The secular consequences of progress diminish both the adaptive, evolutionary value of the will to power and the sociopsychological conditions necessary for its further propagation. The rise of beat philosophy has made this self-limitation so strikingly manifest in our time that it is difficult to escape the conclusion that progress will soon stop in its tracks.

*   *   *

On being confronted with the assertion that progress is now coming to an end, many people seem to dismiss this idea out of hand by pointing out that throughout history there have always been false prophets of limited vision who claimed that after *their* time no further progress could be possible. No doubt, these people think, there was a man in Sumer who said "now that we have invented the wheel, progress has gone about as far as it can go." Quite apart from the logical irrelevance of the failure of past for the bonity of present prophecies, it is not even true that false predictions of the end of progress are of long standing. For the very idea of progress—that history embodies a movement toward a better world—is hardly more than two hundred years old. And hence the first assertion that progress is coming to an end must be of more recent date.

As the chief historian of the idea of progress, J. B. Bury, has set forth, the idea of progress was alien to the Ancients. Their belief in the Golden Age as the beginning of history and its four ever-worse successor ages obviously represents a degenerative rather than a progressive view of history. This degenerative view caused time to be regarded as the enemy of mankind and led to the conservatism of antiquity, which admired stability and deplored change. Medieval Christian philosophy was equally incompatible with the idea of progress, since it held that until Judgment Day, the goal of man's exertions on Earth should be his salvation in another world. On Judgment Day, God will restore the Golden Age, and any do-it-yourself movement of man toward that Golden Age is rendered out of the question by his Original Sin. Nevertheless, medieval anticipation of Judgment Day did introduce one important idea unknown to the Ancients which was to provide the foundation for the later idea of progress: There is a *good* direction to history. The idea of progress still remained undiscovered during the Renaissance. For Renaissance exaltation of classical antiquity as the time when reason had reigned supreme and the arts had attained a pinnacle impossible to reascend could hardly have fostered progressive notions. But one further important element had now emerged: Self-confidence, after its eclipse in the Dark Ages, was restored to reason, and the belief gained ground that there is a purpose to human strife other than salvation beyond the grave. The rise of science in the sixteenth century—particularly the Copernican revolution—tarnished the glory that was Greece and Rome, and by the seventeenth century the first assertions appeared that modern times are actually no worse than antiquity. The French Encyclopedists of the eighteenth century discovered that the cumulative extension of knowledge brought by science causes an amelioration of the human condition and finally, upon the advent of the French Revolution, the idea of progress was given its comprehensive formulation by the Marquis de Condorcet. This idea was to become the central theme of nineteenth-century thought and found reflection in the works of such of its major figures as Karl Marx, Auguste Comte, and John Stuart Mill. In the wake of the publication of Darwin's *Origin of Species* in 1859, the idea of progress was raised to the level of a scientific religion, with Herbert Spencer as its apostle. For since the inexorable processes of evolution are constantly working to improve nature, man's condition obviously partakes in the general movement toward a better world. This optimistic view came to be so widely embraced in the industrialized nations, particularly in America, that the claim that progress could presently come to an end is now widely regarded as outlandish a notion as was in earlier times the claim that the Earth moves around the Sun.

It is now high time that I define progress more precisely. What does movement toward a "better" world really mean? Most people undoubtedly understand a better world to be one of greater *happiness*. But since it is patently impossible to make a meaningful quantitation of happiness—were the medieval serfs more or were they less happy than the denizens of present-day megalopolitan surburbia?—this definition renders belief in progress an act of faith, one not subject to verification or disproof. So this definition is useless for any discussion of the coming end of progress. The definition of progress as "natural" Darwinian evolution toward a "fitter" human condition is equally useless in this connection, because of the tautologous nature of the fitness concept.

It seems that the most meaningful definition of progress can be made from the purview of its very mainspring, namely the will to power. That is, the "better" world is one in which man has a greater power over external events, one in which he has gained a greater dominion over nature, one in which he is economically more secure. This definition makes progress an undeniable historical fact. Furthermore, it makes possible the claim that progress will end, since the assertion that there is to be no further increase in power over external events is meaningful, even if it were untrue. This definition does not, therefore, encompass such wholly internal aspects of the human condition as happiness. Hence, it is a totally amoral view of progress, under which nuclear ballistic missiles definitely represent progress over gunpowder cannonballs, which in turn represent progress over bows and arrows.

In thus focusing on power over external events as the measure of progress, one of its most striking features becomes readily comprehensible. That feature is that progress has proceeded at an ever-accelerating rate. It is commonplace nowadays to present graphs on which diverse indices to the dominion over nature, such as world population, per capita income, speed of travel, world energy consumption, or number of working scientists, are plotted against historical time. Every such graph invariably shows a curve of upward concavity. For the first three thousand years onward from the Middle Kingdom of ancient Egypt, the curve remains practically level at a very low value, then starts a slow rise in Renaissance times, rises still more sharply after the Industrial Revolution, and shoots upward almost vertically in the twentieth century. Another way of appreciating the accelerating character of progress is to note the ever-increasing frequency of discoveries. For instance, if we consider the sources of inanimate energy discovered since fire was first exploited some fifty thousand years ago, we find that forty-five thousand years had to elapse until the next energy source, water power, was harnessed; about thirty-five hundred years after that wind power came into use; within another three hundred years steam power was discovered; the internal combustion engine followed within another century, and nuclear power became available after only forty more years. Or, if we consider the discovery of natural forces since the concept of a natural force was first formulated by the Greeks two thousand years ago, we find that about seventeen hundred years went by before gravity was discovered, followed two hundred years later by electromagnetism, which, in turn, was followed fifty years later by nuclear forces.

These dynamics of progress, and their importance for the understanding of history, were set forth some sixty years ago by Henry Adams, in his "Law of Acceleration." Adams noted that during the nineteenth century the power utilized from the world's coal output, and by inference the rate of progress, doubled every ten years. From the beginning of the fifteenth century until the end of the eighteenth century he judged the doubling period of the rate of progress to be between twenty-five and fifty years. But, Adams pointed out, establishing the actual length of the doubling period is of relatively little importance compared to admitting the fact of acceleration itself. Projecting these dynamics of progress into the future, Adams thought that to an American living in the year 2000, "the 19th century would stand on the same plane with the 4th century—equally childlike—and he would only wonder how both of them, knowing so little, and so weak in force should have done so much." In the twentieth century, however, a new social mind would be required. For "thus far, since five or ten thousand years, the mind had successfully reacted, and nothing yet proved that it would fail to react—but now it would need to jump."

Such kinetics of acceleration are well known to the natural sciences, where they are generally explained in terms of reactions involving an element of *positive* feedback: The farther the reaction has already progressed, the faster its further progress. The growth of a bacterial culture is an example of the most simple of such reactions; since each bacterium in the culture gives rise to two daughter bacteria half an hour after its own birth, the number of bacteria being born per minute is proportional to the number of bacteria that are already present, and hence the rate of growth of the culture doubles every generation. Evidently the element of positive feedback embodied by progress is that the rate at which man can gain more power over the outer world is the greater the more power is already at his disposal. Now if history is viewed as the movement of progress, then history too is accelerating with respect to calendar time.

It is just this acceleration of progress relative to calendar time (and also relative to human physiological time) which accounts for the precise moment in history when the idea of progress finally arose. Throughout antiquity, Middle Ages, and Renaissance, the rate of progress was so slow that the world from which any person departed upon his death was not very different from that into which he had entered upon his birth, even though the fortunes of his own person or community might have undergone some perceptible change. Indeed, what perceptible changes there were, were most often for the worse, such as the ravages of war and pestilence. Some progress *was* taking place during all that time, of course, but it was proceeding so slowly that within living memory things seemed to have either remained as they always had been or deteriorated. And so the hegemony of historicist pessimism in classical times, of medieval hope for redemption only after death, and of Renaissance nostalgia for the glories of Greece and Rome finds its explanation in the impossibility of any personal experience of progress in those far-away days. Toward the end of the eighteenth century, however, an "equivalence point" was finally reached. Now within one life span, the American, French, and Industrial revolutions had so obviously brought about social, political, and economic improvements in the human condition that progress had at last become a matter of personal experience. In the century and a half that has elapsed since that equivalence point, progress has continued to accelerate, so that within the memory of an octogenarian living today the world has changed beyond all recognition.

But this very aspect of *positive* feedback of progress responsible for its continuous acceleration embodies in it an element of

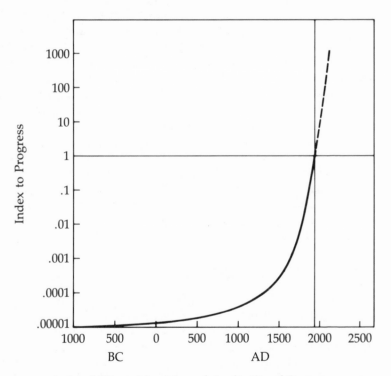

A plot of Henry Adams' Law of Acceleration of Progress.

temporal self-limitation. For since it seems *a priori* evident that there does exist *some* ultimate limit to progress, some bounds to the degree to which man can gain dominion over nature and be economically secure, it follows that this limit is being approached at an ever-faster rate. It is difficult, of course, to quantitate any limit to progress, since the degree to which mankind has gained dominion over nature cannot be expressed in terms of any single parameter. Nevertheless, if one examines one by one the parameters conceivably relevant to estimating the rate of progress, such as world population and energy consumption, per capita income, or speed of travel, one must conclude that none of them is likely ever to exceed some definite bound. And just as, according to Adams, the actual length of the doubling time of progress is of relatively little importance compared to admitting the fact of acceleration itself, so the actual magnitude of the limit to various indices of progress is of relatively little importance compared to admitting the existence of the limit itself. To appreciate this fact, we may consider a semi-logarithmic plot of Adams' Law of Acceleration. On this graph, distances on the vertical axis are proportional to the logarithm of the magnitude of an index to progress, and distances on the horizontal axis are proportional to calendar time. The curve has been drawn according to the doubling times of progress as estimated from Adams' index of world energy consumption figures. The present (A.D. 1960) level of the index has been assigned the arbitrary value of 1.0, representing a 100,000-fold increase over the (1000 B.C.) base line of 0.00001. Extrapolation of this curve into the future shows that any reasonable limit of the parameter under consideration will be reached within a period short with respect to historical time. Thus, even if the present level of 1.0 of the parameter is only one-thousandth of its ultimate limit, that limit would be reached by the year 2160. And, even if the rate of acceleration of the parameter were considerably less than that of world energy consumption, it would still follow from the fact of acceleration that any reasonable limit would be reached within a few centuries. Thus here we perceive a second general conclusion concerning progress: Though progress has occurred in the past, its accelerating kinetics preclude it from being an everlasting feature of human history in the future. Indeed, the dizzy rate at which progress is now proceeding makes it seem very likely that progress must come to a stop soon, perhaps in our lifetime, perhaps in another generation or two.

* * *

Finally we may return briefly to the phenomenon of Bohemians in general and of the beatniks in particular. First, it should have become apparent by now why Bohemianism first arose in the nineteenth century and not earlier. For it was only then that the constant acceleration of progress had reached such a pitch that important social, political, and economic changes had become not only perceptible but were occurring too fast for their general spiritual and cultural assimilation. Consequently, adjustment of mores to the secular consequences of progress could no longer occur at the requisite speed, and there developed an ever-greater discrepancy between the real world and the situation to which these mores were supposed to pertain. This lack of correspondence between the postulated and the actual human condition is most evident to the young simply because *their* psyche developed in a situation which is closer to the present than the childhood environment of their elders. Hence, the most sensitive young people, who can most readily appreciate the changes wrought by progress, become most easily alienated from the society mounted by their elders, and thus tend to emigrate to Bohemia. When the mores of that society finally catch up with the new situation, the now middle-aged ex-Bohemians no longer differ in their attitudes from anyone else; they are "rehabilitated." Meanwhile, of course, progress had caused further, as yet unassimilated changes and moral paradoxes, the appreciation of which gives rise to the next generation of young Bohemians. It is, therefore, reasonable to consider the beatniks as representing another natural link in this alienation chain, which began after the equivalence point in rate of progress and human life span was reached late in the eighteenth century. Second, we now see why beatniks represent a phenomenon of capital importance for understanding the present. It is a mistake to conclude that because of their beat attitude, the beatniks will never

amount to anything, will never, unlike their Bohemian predecessors, create anything of lasting value, and will, therefore,
be without significant effect on the future; here today, gone
tomorrow. On the contrary, by allowing Faustian Man to make
his exit from the stage of history, beat philosophy paved the
way for the profound adjustments in the human psyche that
are necessary if man is to bear living in the Golden Age.

Bibliography

Adams, Henry. *The Education of Henry Adams* (Chapters 23 and 24).
Boston: Massachusetts Historical Society, 1918.

Aiken, H. D. *The Age of Ideology: The 19th Century Philosophers.* New
York: Mentor Books, 1956.

Bury, J.B. *The Idea of Progress.* New York: MacMillan, 1932. Reprinted by
Dover Publications, New York, 1955.

Muscatine, C. (Chm.). *Education at Berkeley: A Report of the Select Committee on Education.* Berkeley: Univ. California, 1966.

Ortega y Gasset, J. *The Revolt of the Masses.* New York: Mentor Books,
1950.

Parry, A. *Garrets and Pretenders: A History of Bohemianism in America.*
New York: Dover Publications, 1960.

The Berlin Philharmonic. Architect: Hans Scharoun.
[Courtesy Infobild, Berlin.]

2

THE END
OF THE
ARTS AND
SCIENCES

[*1969*]

In the preceding chapter I adduced two independent arguments which lead me to the belief that progress will stop in our time. One argument, psychological in kind, asserted that accession to economic security, a secular consequence of progress, will ultimately brake progress because it is inimical to the paragenetic transmission of the will to power. The appearance of the beat generation was interpreted in the light of this argument to mean that the negative feedback of progress on the will to power has by now made its effects felt on a massive scale. The other argument, kinetic in kind, pertained to the constantly accelerating rate of progress. For if, as seems *a priori* reasonable, there are *some* limits to progress, then these limits are being approached at an ever-faster speed. The rate of progress has now become so fast—such fantastic changes in the human condition have taken place within living memory—that it seems difficult to imagine any limit to progress so far beyond what has already been accomplished that this limit will not be reached soon. Of these two arguments, the second is patently the weaker, because it depends on the impressionistic finding that the present rate of progress *is*, in fact, fast with respect to the approach of its ultimate limits occurring in a human life span. I now want to adduce a third, an entirely independent argument which shows that ultimate limits are already in view in what is generally considered to be the most important indices to progress. These indices are the arts and sciences, in which sublimated will to power and exertion of Faustian Man have found their highest expression.

It is a commonplace reaction to the present situation in the arts to sense that there seems to be something drastically wrong. Such contemporary manifestations as action painting, Pop art, and chance music are causing widespread apprehension about the state of art. And this apprehension appears to exist not only among the general public but also among an important sector of the artistic community itself. (Lewis Mumford, for example, wrote recently that "the fashionable oppish and poppish forms of non-art today bear as much resemblance to . . . exuberant creativity . . . as the noise of a premeditated fart bears to a trumpet voluntary of Purcell.") Many people feel that art has somehow turned into a dead-end street and that for there to be *any* future an escape must be found from the present direction. The theme I want to develop now is that there is no such escape, inasmuch as these latter-day bizarre art forms have but followed in natural succession from the masterworks of the past. Instead of making a wrong turn, art has merely traveled down (or, rather up) a one-way street since its beginnings in the remote prehistoric past. In order to make this point, I want to draw attention to one obvious historical trend which provides the directional arrow for traffic along that one-way street: As artistic evolution unfolds, the artist is being freed more and more from strict canons governing the method of working his medium of creative expression. The end result of this evolution has been that, finally, in our time, the artist's liberation has been almost total. However, the artist's accession to near-total freedom of expression now presents very great cognitive difficulties for the appreciation of his work: The absence of recognizable canons reduces his act of creation to near-randomness for the perceiver. In other words, artistic evolution along the one-way street to freedom embodies an element of self-limitation. The greater the freedom already attained and hence the closer the approach to the random of any artistic style for the percipient, the less possible for any successor style to seem significantly different from its predecessor. I will now try to present a brief summary of what I understand to be the information-theoretical and psychological reasons underlying this dominant trend in artistic evolution. This argument is gleaned from the works of such writers as Suzanne K. Langer, Wylie Sypher, Leslie A. Fiedler, and especially Leonard B. Meyer. Furthermore, I will attempt to show later that a similar element of cognitive self-limitation is apparently involved also in the progress of the sciences, a limitation the existence of which has only recently come to light.

To begin this discussion, I will state the traditionalist assertion that both the arts and the sciences are activities that endeavor to discover and to communicate truths about the world. These endeavors became possible at that remote but most critical stage in man's psychic evolution when he turned into a semantic animal, that is, when he got hold of the powerful idea of the symbolic representation of events. The domain to which art addresses itself is the inner, subjective world of emotions. Artistic statements, therefore, pertain mainly to relations between private events of affective significance. The domain of the sciences, in contrast, is the outer, objective world of physical phenomena. Scientific statements, therefore, pertain mainly to relations between or among public events. This dichotomy of domain does not mean that for the percipient an artistic statement is necessarily devoid of objective significance (a Canaletto painting, for instance, provides information about the public event that was Venice), nor that a scientific statement is necessarily devoid of affective significance (for instance, a great discovery by another scientist in one's own field is usually a source of deep disappointment, and rarely of great joy, as will be seen in Chapter 4). In any case, from this traditionalist standpoint information and the perception of *meaning* in that information is the central content of both arts and sciences. Hence when we speak of progress in the arts and sciences we can really refer to only one thing, namely that progress is taking place as long as the sum total of meaningful artistic and scientific statements waxes. And thus progress in the arts and sciences would be reaching its end when it has become more and more difficult to continue adding to that accumulated capital of meaningful statements.

In the beginning, the origins of art were not yet art. For example, music probably took its first roots in the organization of sounds, vocal and percussive, for rhythmatization of work and ritual and for nervous excitation. At this stage music was not yet art—no more than primitive grunts and cries were language—because music still lacked its semantic function. Only upon the gradual development of symbolic use of musical form did music become art. In the dichotomous scheme of artistic and scientific domains, music appears to be the "purest" of the arts, in that it has the least to say about the outer world and hence shows the least overlap with the sciences. The thematic content of music has very few "outer" models—bird calls, hoofbeats, and thunderstorms among them—compared to, say, the vast variety of visual impressions or prototype situations available as models for the visual or dramatic arts. Indeed, music can easily dispense with outer models, to which it could never do justice anyway. Thus the content of music is necessarily more purely affective than that of any other art form; its statements pertain almost exclusively to inner events. Musical symbolism is able to dispense with natural models, because, according to Mrs. Langer, "the forms of human feeling are much more congruent with musical forms than with the forms of language; music can *reveal* the nature of feelings with a detail and truth that language cannot approach." Hence music conveys the unspeakable; it is "incommensurable with language, and even with presentational symbols like images and gestures." "Program" music, which does attempt to represent outer events, appears to be an exception that proves the rule, in that program music is generally accorded a rather low artistic merit.

How does symbolic meaning arise from the temporal sequence of tones that we perceive? According to Meyer, "musical meaning arises when an antecedent situation [of tone sequences], requiring an estimate [by the listener] of probable modes of pattern continuation, produces uncertainty about the temporal-tonal nature of the expected consequent." This definition derives from a general consideration of the nature of information. The amount of information embodied in any event is the higher the greater the number of alternative events which the percipient expected would occur given the antecedent situation. If that situation is so highly structured that the percipient's expectation of occurrence of the event is very high, the information content of the event is low. But the *meaning* of the information provided by the event derives from the evaluation of that information in respect to past and future events. That is, for an event to have meaning, its occurrence must not only have been uncertain but it must be capable also of modifying the probabilistic appreciation of the consequences of the earlier antecedent situation. Thus as a meaningful piece of music unfolds, the listener is constantly modifying his expectations of what he will hear next on the basis of what he has already heard. His final probabilistic connection of the entire tone sequence is the musical form the listener has recognized, the structure he has perceived.

Now in the listener's process of estimating the probable modes of pattern continuation from the antecedent tone sequence, there enter not only the informational feedback evaluations which he has inferred from what he has already heard of the musical composition to which he is presently listening but also the statistical rules governing possible tone sequences which he has abstracted from his previous listening experience of other, similar compositions. And these statistical rules governing possible tone sequences are nothing other than the *style* in which the piece of music has been composed. That is, the listener can actually make an estimate of the probable modes of pattern continuation only if he has some awareness of the stylistic canon under which the composer operated. And here we reach an important point, namely that for the composition to be most meaningful, there exists for any listener an optimization of rigidity of the stylistic canon. If, on the one hand, the canon is *too rigid,* the uncertainty about the temporal-tonal sequence to come is very small and the redundancy of its information very high. Hence the listener hears mainly what he was certain to hear all along; the rate at which information is conveyed to him is very low; he has little reason to modify his probabilistic appreciation of the antecedents; he has learned very little; the piece is, therefore, nearly without any meaning. If, on the other hand, the canon is *too lax*, the uncertainty about the temporal-tonal sequence to come is very great, and the redundancy of its information very low. Hence the rate at which information is conveyed to the listener will, therefore, be very high. But the speed at which new information impinges on him may exceed his "channel capacity," that is, he will not be able to evaluate that information fast enough to abstract a probabilistic appreciation of the antecedents, particularly if the paucity of redundant information does not allow him to test the validity of his inferences. Here too the piece is, therefore, nearly without meaning. Thus for a listener to perceive a significant structure in a musical composition, it must present him with a temporal-tonal sequence which is neither too certain nor too uncertain. That is, the freedom of the style of composition must match the musical sophistication of the listener.

From this information-theoretical purview, creativity in musical composition evidently represents the generation of meaningful new structural patterns. But here we can perceive the reasons for the historical trend for progressive relaxation of stylistic canon. The point of optimum rigidity of stylistic canon for communication of significant meaning *must* evidently move in the direction of greater freedom as listener sophistication waxes thanks to the accumulated capital of previously created significant structures. At the beginning when music consisted only of rhythmic chants, drumbeats, and vocal imitation of natural sounds, stylistic canon was at its most rigid; there existed almost no compositional freedom; listener sophistication was minimal. To create meaningful new musical patterns, it was necessary to relax the canon a little, but not too much, for instance, by allowing for the inclusion of a few unnatural tone sequences. Presently, this had two consequences: Listener sophistication rose, and the possibility to create new meaningful patterns became exhausted. A further relaxation of canon now occurred (that is, a new style was created), composition of new significant patterns became possible, and listener sophistication rose further. And so it went from antiquity through the Middle Ages, Renaissance, Baroque, Romantic, and Impressionist periods down to contemporary atonal music: the appearance of a new, somewhat less rigid style engendering a mounting listener sophistication, followed later by exhaustion of the possibilities for significant new creations in that style and resulting finally in the appearance of a new, yet freer, successor style. It might be noted here in passing that the freer of two styles is not necessarily that which operates with fewer and/or less complex rules. Obviously, a small number of very simple rules can give rise to a very rigid and highly redundant style.

It would thus follow that there has occurred an evolution of musical style from its primitive origins toward an ever-higher level of sophistication. The inference of such an evolution does not rest on any teleological view of man or music necessarily evolving toward a higher goal; it devolves merely from a recognition of the information-theoretical basis of their interaction. The kinetics of that evolution, furthermore, manifest the same constant acceleration which we have already noted to have held for progress in general. The styles of classical antiquity and Middle Ages lasted for many centuries, those of the Renaissance for a century or two, those of the Baroque period and Romanticism for many decades, those of Impressionism for a decade or two, and finally contemporary styles succeed one another within a matter of a few years. This acceleration might reflect, in part, the constantly greater number of working composers corresponding to the rise in world population. Their ever-waxing aggregate activity exhausts ever more rapidly the possibilities for significant creations within any given style, though sheer greater number alone does not, of course, guarantee a heightened rate of creativity. Thus one man, such as J. S. Bach, has probably done more to exhaust the possibilities of *his* style than the aggregate effort of all of his lesser contemporaries. And thus a more important reason for the acceleration of stylistic evolution is probably technological progress in the media of musical communication. For instance, the invention of musical notation must have been a very important first step, which finally secured the accumulation of musical capital against the vagaries of human memory. The invention of printing then allowed a wide distribution of that capital to potential performers, and finally the advent of phonograph, radio, L.P. record, and tape recorder resulted in the rapid dissemination of new compositions among a vast audience. Thus listener sophistication could rise at an ever-greater rate, allowing in turn for an ever-faster stylistic evolution.

Serial music pioneered by Arnold Schönberg represents a late but by no means final stage in this evolution. The composer has now been freed of any restraints imposed on him by the traditional dictates of melody and harmony, but his freedom is not yet total. The older canons have been replaced by the laxer rules of the twelve-tone row, but rules still do exist. However, these laxer rules have now gone so far toward reducing the redundancy of information in the temporal-tone sequence that "learning" serial music already presents a difficult perceptive problem; having previously learned one piece of serial music is of relatively little help in learning the next piece, other than the general training such learning would have provided in mastering a difficult task of musical cognition. But the final stages of this evolutionary process have now been reached with the experimental music of such composers as John Cage. For here almost all rules that would allow communication to the listener of a musical structure have been abandoned. In one type of such experimental music the temporal-tonal sequence is purposely generated by pure chance, either by the composer in writing or by the performer in reading the score, so that the form is intentionally random. In another type, the composer writes intuitively without consciously attempting to develop any particular idea or to reach any ultimate goal. Thus the listener is left to his own devices, to make of the music what he will. The structure he perceives in the piece, if any, is entirely dependent on his own personality, much as his interpretation of an inkblot in the Rorschach test also depends on his personality. Thus with this development, music as an art which endeavors to discover and communicate truths about the world *has* reached the end of the line.

What, then, do these composers of experimental music have in mind? What are they trying to do? To fathom the nature of their activity, it is necesary to appreciate that the view of the world of these latter-day artists is radically different from that traditionally associated with rational thought. This view, which Meyer has called *transcendentalism,* shows strong affinities to the precepts of Zen Buddhism, in that the transcendentalist believes that concrete, particular sense experiences are the only truths to be found in the world. Any attempt to construct a reality by inferring imaginary causal relations between or among these sense experiences obscures rather than reveals the essential truth of existence, namely that every fact of the universe is unique. It becomes apparent at once that to anyone holding such a belief the very idea is anathema that the meaning of a piece of music for the listener devolves from the structure he perceives in the probabilistic connections of its temporal-tonal sequence. Instead, for a transcendentalist the music is just *there*, and any analytical cerebrations only interfere with its experience as a primary fact. Art and nature thus merge into one: there is no qualitative experiential distinction between listening to the sound of music and the noise of nature. Thus the transcendentalist composer of experimental music not only does not add to the accumulated capital of meaningful statements about the world, but nothing could be farther from his mind than intending to do so. His sole purpose is to add to the sum total of unique sense experiences of his listeners.

* * *

Our discussion of the arts has, thus far, been focused on the evolution of music, without consideration of the fate of such other important domains as painting, literature, poetry, and drama. Since this essay has already overtaxed my competence in the aesthetic realm, I shall not attempt to reproduce an equivalent information-theoretical argument to account for the historical trend toward greater freedom for artists working in other, nontonal media. But I think it is fair to say that essentially the same process of exhaustion of the possibility for creating significant meanings within a style of given rigidity adapted to the level of audience sophistication, followed by invention of a slightly less rigid style and repetition of the audience education-style exhaustion dialectic, must have been at work in the nontonal arts as well. In any case, by now nearly all of these other art forms too seem to have reached what appear to be terminal or near-terminal stages in their development formally equivalent to that of experimental music. That is, the nontonal arts have now evolved styles by means of which meaningful communication in the information-theoretical sense between artist and percipient is neither possible nor intended. As far as the visual arts are concerned, this terminal genre is represented by such styles as action painting, as practiced by painters who drip or splash paints on their canvas, and by Pop art, as exemplified by the eclectic collages of "found" objects and the facsimiles of Campbell soup cans and comic strips. As Sypher has pointed out, the unifying characteristic of these styles is the anonymity of the artist whose self finds no reflection in his works. And no more than the experimental composer does the action painter and the Pop artist fashion his works as new, meaningful statements about the world. He merely adds to the experiential repertoire of his audience, which is to make of these works what it will. Similarly evident cases of terminal stages in art seem to have been reached in drama and literature. Here, no subtle information-theoretical analysis is required to show how meaning arises from the medium of drama and literature, for the "language" of drama and literature is obviously language itself. But since playwrights and writers have not generally manipulated grammatical rules for their own purposes, drama and literature has not undergone any information-theoretical development in the use of the medium of communication comparable to that which occurred in music and the visual arts. Instead, the possibilities of using language as an artistic medium simply seem to have been used up. In drama, this exhaustion finds reflection in the theater of the absurd, particularly in the works of Eugene Ionesco. In our time, so Ionesco realized, all verbal language has become a cliché and hence is no longer suitable for communicating matters of affective significance. And so the dramatist of the absurd, like the experimental composer, action painter, and Pop artist, has abandoned the notion of conceiving his work as a message. The characters of the theater of the absurd mouth meaningless words, lack any identity, and engage in actions that are not causally connected, that is, do not weave into a plot. The actor's main function is to be on stage, to be *there*. In literature, the end of the novel has become manifest with the appearance of the works of such writers as Alain Robbe-Grillet and William Burroughs. In their antinovels, all semblance of organization has disappeared. There are no rational connections between individual sentences and paragraphs, there are no characters, there is no story. Fiedler points out that the rise of the now dying novel in the nineteenth century was itself a big step toward the end of literature. From the purview of eighteenth-century epic poetry, the novel was already antiliterature, because, according to Fiedler, "while pretending to meet formal standards of literature, it is actually engaged in smuggling into the republic of letters extraliterary satisfactions. It not merely instructs and delights and moves, but also embodies the myths of a society, serves as scriptures of an underground religion, and these latter functions, unlike the former ones, depend not at all on any particular form, but can be indifferently discharged by stained-glass windows, comic strips, ballads, and movies. Yet it is precisely this cultural *ambiguity* of the novel which made it for so long popular on so many levels, at the same time creating those tensions and contradictions by virtue of which it is presently dying."

Indeed, it is a striking fact that even in so workaday a branch of art and one so close to science as is architecture, a hint of a stylistic end has now become perceptible. For here too an element of the random in design has lately come to the fore. Naturally, in his efforts as an engineer, namely in designing a building that does not collapse and, hopefully, serves its specified function, the architect is constrained to obey many rather strict rules. But in his efforts as an artist, as a creator of aesthetic truths, it would appear that the architect's channels of significant communication are also nearing their limits. For instance, upon beholding so random a structure as Hans Scharoun's new Berlin Philharmonic, it is difficult to escape the feeling that one is beholding a work belonging to some final phase of architectural style. It is difficult to imagine any other building of a significantly different form. The Philharmonic is just *there*. The conceivable development of revolutionary new building materials or techniques, which had always brought about radical stylistic changes in the past, seems unlikely to affect this conclusion very seriously, except insofar as it might give the architect still greater freedom to proceed to the design of even more random, ultimately nonbuildings.

Possibly the cinema, being a medium of such recent invention, is one of the few art forms whose end is not yet so clearly in sight. Its possibilities do not seem to have been so fully exhausted that radically new styles are impossible to imagine. Perhaps it is for this reason that the cinema seems to have gained ascendancy over the theater in the recent past.

* * *

How, then, is the future of art to be envisaged if stylistic evolution has now reached the end of the line? Meyer is of the opinion that "the coming epoch (if, indeed we are not already in it) will be a period of *stylistic stasis*, a period characterized not by the linear, accumulative development of a single fundamental style, but by the coexistence of a multiplicity of quite different styles in a fluctuating and dynamic steady-state. . . . In music, for instance, tonal and non-tonal, [chance] and serialized techniques, electronic and improvised means will all continue to be employed. Similarly in the visual arts, current styles and movements—abstract expressionism and surrealism, representational and Op art, kinetic sculpture and magic realism, Pop and non-objective art—will all find partisans and supporters. Though schools and techniques are less clearly defined in literature, present attitudes and tendencies—the 'objective' novel, the theater of the absurd, as well as more traditional manners and means—will, I suspect, persist." Thus Meyer envisages that in addition to those artists who work in latter-day, transcendentalist styles in which meaningful communication between artist and audience is no longer possible or intended, there will continue to be other artists who will persist in the use of the older, semantic styles. Though the former will not, of course, add to the accumulated capital of meaningful statements, the latter might continue to do so indefinitely. And Meyer says he knows of "no theoretical or practical reason why a talented, well-trained contemporary composer could not write, say, an excellent concerto grosso in the manner of the late Baroque. And though, unless he were a man of genius, the composition would surely fall far short of the work of Bach, it might easily compare favorably in interest and quality with countless works of lesser Baroque composers." According to Meyer, no such anachronistic use of past styles was made until the present time, because it was considered corrupt, contemptible, and dishonest from the purview of our cultural beliefs about originality and creation, causation and history. But, so continues Meyer, the abandonment of these beliefs and their replacement by the philosophy of transcendentalism will remove all barriers to the warming over of the styles of past epochs.

Nevertheless, it seems to me unlikely that the future use of past styles in the coming epoch of stylistic stasis will permit much further progress in the arts. From Meyer's argument it would follow that the very reason why the Baroque style came to be abandoned was that Bach had simply exhausted its creative possibilities. And, as T. S. Eliot wrote in a passage quoted by Meyer, "when a great poet has lived, certain things have been done once and for all and cannot be achieved again." Hence the talented composer of the future would seem to be ill-advised to choose the style of the late Baroque if he had anything significantly original to convey. Of course, in this case he would not be a transcendentalist anyway and might have scruples about resorting to stylistic atavism. But if the talented composer of the future *is* a transcendentalist and thus feels free to use any style, past or present, then he would not, by definition, compose in a meaningful way. That is, *his* Baroque concerto grosso would be semantically as nugatory as the Pop artist's Campbell soup can.

*   *   *

Whereas presentiments of a coming end to art have now become a commonplace, the possibility of an end to science is much more rarely bruited. One remembers, of course, that often-told episode of how some *fin-de-siècle* physicists thought that physics was nearing its end. The grievous error of those people in the light of the then imminent advent of quantum and relativity theories has taught later generations the lesson that one can never know what unexpected scientific discovery is just about to show up. I must admit that this hortatory tale *ought* to give anyone pause for thought who predicts an end to the sciences. But as Meyer, fully aware of *his* exposed position as the prophet of artistic stasis, points out in reference to earlier false predictions of a coming end of artistic evolution, no one believed the boy who cried "wolf" wrongly once too often, but then the wolf finally *did* come. And so even though nowadays nearly all scientists still seem to envisage an unlimited progress of our knowledge of nature, I shall now set forth some arguments from which it can be concluded that for the sciences, as for the arts, an end is in sight.

First, I want to consider briefly a possible socioeconomic limitation to science. Since the nineteenth century it has become generally recognized that the fruits of scientific research lie at the roots of economic progress and that they are responsible for man's gaining ever-greater dominion over hostile nature. Indeed, it has finally dawned on the governments of the technically advanced nations that support of scientific research has so far paid the highest rate of return of any social investment. Accordingly, an ever-increasing fraction of the gross national products of these nations has been consecrated to the sciences, which in turn have become more and more expensive to conduct. But as that quest for dominion over hostile nature is nearing its goal, as technological advances made possible by the application of the results of scientific research vanquish all threats posed to human survival by hunger, cold, and disease, further scientific research appears to have arrived at the point of ever-decreasing utility. Thus it seems possible that there could occur a waning of the present high social interest in supporting the sciences. This argument might, however, lose its validity if upon the advent of what Herman Kahn calls the "post-economic" age, the sciences are still a going concern. For by that time technological progress might have brought about a virtually infinite gross national product, a condition under which utilitarian considerations for deciding upon the magnitude of social support for various activities will have lost their relevance.

Second, and more importantly, I want to consider what I believe to be intrinsic limits to the sciences, limits to the accumulation of meaningful statements about the events of the outer world. I think everyone will readily agree that there are *some* scientific disciplines which, by reason of the phenomena to which they purport to address themselves, are *bounded*. Geography, for instance, is bounded because its goal of describing the features of the Earth is clearly limited. Even if the totality of the vast number of extant topographic and demographic details can *never* be described, it seems evident nevertheless that only a limited number of significant relations can ultimately be abstracted from these details. And, as I tried to show in *The Coming of the Golden Age*, genetics is not only bounded, but its goal of understanding the mechanism of transmission of hereditary information *has*, in fact, been all but reached. Indeed, and here I will probably part company with some who might have granted me the preceding example, even such much more broadly conceived scientific taxa as chemistry and biology are also bounded. For in the last analysis, there is immanent in their aim to understand the behavior of molecules and of "living" molecular aggregates a definite, circumscribed goal. Thus, though the total number of possible chemical molecules is very great and the variety of reactions they can undergo vast, the goal of chemistry of understanding the principles governing the behavior of such molecules is, like the goal of geography, clearly limited. As far as biology is concerned, after the triumph of molecular genetics there now seem to remain only three deep problems yet to be solved: the origin of life, the mechanism of cellular differentiation, and the functional basis of the higher nervous system. In my view, the insights offered by the central dogma of molecular genetics will presently provide the keys for solving also these last problems. And, considering the host of biologists now standing ready to do battle and the vast armory of experimental hardware at its disposal, origin of life, differentiation, and the nervous system cannot help but soon suffer the fate that was accorded to heredity in these last twenty years. I do not, of course, include the solution of the mechanism of consciousness among these sanguine predictions, since its epistemological aspects both posit it as *the* central philosophical problem of life and also place it beyond the realm of scientific research.

Thus the domain of investigation of a bounded scientific discipline may well present a vast and practically inexhaustible number of events for study. But the discipline is bounded all the same because its goal is in view. The awareness of this intellectual horizon embodies in it a yardstick for value, since the greatness of a scientific insight can be measured in terms of the magnitude of the forward leap toward the attainment of that goal that it represents. Hence there is immanent in the evolution of a bounded scientific discipline a point of diminishing returns; after the great insights have been made and brought the discipline close to its goal, further efforts are necessarily of ever-decreasing significance.

There is at least one scientific discipline, however, which appears to be *open-ended,* namely physics, or the science of matter. Whereas the goals of the bounded disciplines are, in the last analysis, defined in terms of physical concepts, the goal of physics of understanding matter must necessarily remain undefined and hence hidden from view. In other words, it is diffucult to envision a set of statements that would "explain" the nature of matter. For such an explanation can be provided only by *meta*physics, in the true sense of that term. Thus there might be no limit to the significant statements that physics can be expected to provide. Indeed, physics might yet generate an unlimited number of bounded subdisciplines (as, say, it generated mechanics in the past) for the sciences of the future. But even though physics is, in principle, open-ended, it too can be expected to encounter limitations in practice. As has been pointed out by Pierre Auger, there are purely physical limits to physics because of man's own boundaries of time and energy. These limits render forever impossible research projects that involve observing events in regions of the universe more than ten or fifteen billion light-years distant, traveling very far beyond the domain of our solar system, or generating particles with kinetic energies approaching those of highly energetic cosmic rays.

Furthermore, the very open-endedness of physics seems to be bringing to it a heuristic limitation, paradoxical as this assertion may seem. Insofar as I am able to judge, the frontier disciplines at the two open ends of physics, cosmology and high-energy physics, seem to be moving rapidly toward a state in which it is becoming progressively less clear what it actually *is* that one is ultimately trying to find out. What, actually, would it *mean* if one understood the origin of the universe? And what would it mean if one had finally found the most fundamental of the fundamental particles? Thus the pursuit of an open-ended science also seems to embody a point of diminishing intellectual return. That point is reached with the realization that its goal turns out to be hidden in an endless, and ultimately tiresome succession of Chinese boxes.

For the purpose of this discussion, mathematics belongs to a special category, in that it appears to occupy a position intermediate between the arts and the sciences. Since the domain of mathematics is logic, it straddles the inner world of private events in which logic arises and the outer world of public events to which logic is applied. It is my understanding that with the appearance of Gödel's theorem some thirty-five years ago, mathematics has certainly become open-ended. For that theorem has shown that any set of axioms of complexity comparable to that which embodies our concept of number will generate some propositions whose truth or falsity cannot be demonstrated, except by making that set part of a larger axiomatic system. That larger system, however, will in turn generate further undecidable propositions. It would not, therefore, surprise me to learn that mathematics too would soon reach a point of diminishing returns.

Auger considers also the possibility that there are mental limits to physics because of man's boundaries of intellect. Auger asks "whether there is not a natural limit to the range of abstraction and complexity which can be covered by human thought, and in particular mathematical thought? The number of nerve cells in the brain, though considerable, is not infinite, nor is the number of connections established between them." I find that this is an important point, even though the way Auger has phrased it might make it seem as though he has overlooked the obvious possibility that the articulation of brain and computer might provide an indefinitely large extension of the number of "nerve cells" available for thought. But there would seem to exist an intellectual limit to physics which is most unlikely to be transcended by any future recourse to auxiliary logical hardware provided by computers. This limit devolves from the circumstance that the fundamental, and I suppose innate, human epistemological concepts, such as reality and causality, arise from a dialectic between the facts of life of our infantile environment and the genetically determined wiring diagram of our brain. Evolution selected this brain (and the bent for ontogenetic development of its innate epistemology) for the capacity to deal "successfully" with superficial, everyday phenomena, but it was not selected for handling such deeper problems as the nature of matter or of cosmos. Or, stating this in a different way, our innate concepts represent an axiomatic system, which, according to Gödel's theorem, contains open-ended propositions. When we encounter such propositions and try to deal with them by tampering with our innate axioms, we pay for the gain in logical coherence with a loss of psychic meaning. For instance, though the replacement of deterministic by probabilistic causality in the consideration of subatomic phenomena has made possible their successful theoretical formulation, the results achieved seem to do violence to common sense.

Now, the obstacle to scientific progress posed by common sense is, like the lack of imagination of the false prophets of the end of physics, the subject of another traditional homily preached in the nurseries of natural philosophy. Common sense, so it is explained to the student, told men that the Earth is flat, that the Sun moves around the Earth, and that forces cannot act at a distance. So, common sense long prevented the recognition what we now know to be true and most readily accept. In other words, yesterday's nonsense may become today's common sense. But I think that this canonical view of the obstructive role of common sense in the history of science is rather superficial since it does not reckon with the psychological consequences of this evolution. First, the present-day ideas that the Earth is round and that it moves around the Sun through the intervention of forces acting at a distance are not really developed as part of his common sense by the growing child through use of his innate epistemological axioms in dealing with the outer world of his infantile environment. Instead, these unnatural abstractions are imposed on him at an intellectually more mature age by adults. Second, it is my belief that every such act of countermanding common sense produces a quantum of alienation from reality, or engenders a partial erosion of the "reality principle" (to which I shall return in the next chapter). Thus we may perceive another internal contradiction in science: the innate axioms on which our brain bases its cognition of the outer world and from which springs common sense suffer ever-greater violation as the evolution of physical research unfolds. This intellectual process causes in turn a progressive estrangement from the reality of that outer world, loss of psychic meaning of the insights gained into its operation, and hence weakening of the intensity of interest in probing further into its phenomena.

* * *

What about the "young" social sciences? Are they not the sciences of the future, for whose development there is now the most pressing need? Surely, there remain to be discovered many fundamental principles of economics and sociology whose application will finally allow man to control not only hostile nature but also his intercourse with fellow human beings. But here we encounter the third and what, for the purposes of this discussion, I consider to be a most important obstacle to further progress in the sciences. This obstacle was, to my knowledge, first recognized by the mathematician Benoit Mandelbrot a few years ago when he began to attempt a statistical analysis of some econometric time series, such as fluctuations in cotton prices. In the course of this analysis Mandelbrot developed an epistemological argument whose applicability transcends economics and which draws attention to a rather more fundamental barrier to the easy forward march of our capacity to discover new laws in both natural and social sciences. This argument has some considerable affinity to the preceding analysis of perception of meaning in music; indeed, it might be useful for this discussion to consider science as the perception of the music of nature. The following is a rather superficial summary statement of what I understand to be Mandelbrot's general argument and his main conclusions.

Let us recall, first of all, that science—that is, the effort to abstract causal relations from observable public events of the outer world—is by its very nature a statistical endeavor. The scientist thinks he recognizes some common denominator, or structure, in an ensemble of events, infers these events to be related, and then attempts to derive a "law" explaining the cause of their relation. An event that is unique, or at least that aspect of an event which makes it unique, cannot therefore be the subject of scientific investigation. For an ensemble of unique events *has* no common denominator, and there is nothing in it to explain; such events are *random*, and the observer perceives them as noise. Now since every real event incorporates *some* element of uniqueness, every ensemble of real events contains some noise. And so the basic problem of scientific investigation is to recognize a significant structure of an ensemble of events above its inevitable background noise. This perceptual problem is thus formally analogous to recognizing the meaning of the tone sequence in nontranscendentalist music. It is, in fact, but another instance of the fundamental information-theoretical problem of distinguishing signal from noise in any kind of communication. Hence the lower the background noise of a natural phenomenon—that is, the smaller the role of the uniqueness of its constituent events in the over-all picture—the more unambiguous is its structure. And just as listener sophistication rose in the evolution of musical styles of less and less structured temporal-tonal sequences, so did observer sophistication rise in the evolution of scientific analyses of less and less structured phenomena. Thus, most of the natural phenomena for which successful scientific theories had been worked out prior to about one hundred years ago are relatively noise-free. Such phenomena were explained in terms of *deterministic* laws, which assert that a given set of initial conditions (antecedent situation) can lead to one and only one final state (consequent). But toward the end of the nineteenth century the methods of mathematical statistics came to be trained on previously inscrutable physical phenomena involving an appreciable element of noise. This development gave rise to the appearance of *indeterministic* laws of physics, such as the kinetic theory of gases and quantum mechanics. These indeterministic laws envisage that a given set of initial conditions can lead to several alternative final states. An indeterministic law is not devoid of predictive value, however, because to each of the several alternative final states there is assigned a probability of its realization. Indeed, a deterministic law can be regarded as a limiting case of a more general indeterministic law in which the chance of the occurrence of *one* of the alternative final states approaches certainty.

[Here it might be well to give a little credit to the erstwhile benighted false prophets of the end of physics; at least they seemed to have correctly sensed the end of *deterministic* physics in their time.] The conventional acid test of the validity of both deterministic and indeterministic laws is the realization of their predictions in future observations. If the predictions *are* realized, then the structure which the observer believes to have perceived in the original phenomenon can be considered to have been real.

Now Mandelbrot asserts that science is presently at the threshold of what he calls a *second stage of indeterminism*, in that many of those noisy phenomena which continue to elude successful theoretical understanding will not only be inaccessible to analysis by old-style deterministic theories, but might prove refractory also to formulation in terms of latter-day, or first-stage indeterministic theories. In making this point, Mandelbrot draws attention to the statistical character of the noise presented by the random aspect of an ensemble of natural events, or the *spontaneous activity of the system.* It is the nature of the spontaneous activity of a system which is of the utmost importance for its cognition. In almost all systems for which it *has* so far been possible to make successful first-stage indeterministic scientific theories, the spontaneous activity displays a statistical distribution such that the mean value of a series of observations converges rapidly toward a limit. That limit can be subjected to analysis of the classical deterministic type. For instance, in the successful kinetic theory of gases, the spontaneous activity of a gas satisfies this condition. Here the energy of individual molecules is subject to a very wide variation (thermal unrest), but the mean energy per molecule converges to a limit and is, therefore, for all practical purposes determined. But many of the phenomena for which it has *not* been possible to make successful scientific theories so far turn out to possess a spontaneous activity which displays quite a different distribution. For such phenomena the mean value of a series of observations converges only very slowly, or not at all, toward a limit. And here, according to Mandelbrot, it is very much more difficult to ascertain whether any structure the observer believes to have perceived is real, or merely a figment of his imagination. To illustrate this point, Mandelbrot cites the record of a century-long coin-tossing game between Peter and Paul, reproduced here. If we focus our attention on the points where the fortunes of Peter and Paul are equal (that is, where the record crosses the horizontal line), then we observe that the density distribution of those points is extremely irregular. In particular, it is apparent that the relative variability in the number of such crossings per time interval is not decreased by considering longer and longer intervals. In such a record, a wealth of detail and structure can be perceived by an interested observer (say, a gambler). But any perceived structure is evidently a mere illusion of the observer's brain that has no bearing on the random mechanism which actually generated the record and which will generate future events.

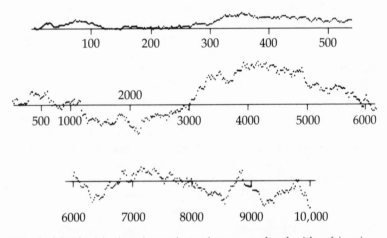

Record of Paul's winnings in a coin-tossing game, played with a fair coin. Zero crossings seem to be strongly clustered, although intervals between crossings are obviously statistically independent. To appreciate fully the extent of the apparent clustering in this figure, note that the units of time used on the second and third lines equal twenty plays. Hence the second and third lines lack detail, and each of the corresponding zero crossings is actually a cluster, or a cluster of clusters. For example, the details of the clusters around the time 200 can be clearly read on the first line, which uses a unit of time equal to two plays. [*From W. Feller,* An Introduction to Probability Theory and its Applications, *second edition. John Wiley & Sons, New York (1957).*]

Mandelbrot suggests the Gedankenexperiment of having an explorer bring back Peter and Paul's coin-tossing record as the topographical cross section of a hitherto unknown part of the world, in which all the regions below the bold horizontal lines are under water. Evidently, this record manifests such "typical" geographical features as "oceans," "islands," "archipelagos," and "lakes." The question now is how to decide whether the formation of this topography was due to cause or to chance. Evidently, any such decision will be of the utmost difficulty. This hypothetical problem is, in fact, close to the real situation, in that the variation in the size of terrestrial islands follows the same kind of statistical distribution as does the variation in the distances between zero crossings of the coin-tossing record. This kind of distribution is called "Pareto" distribution, after the turn-of-the-century Italian economist who first observed it in the distribution of incomes. Indeed, many other geophysical, meteorological, and astrophysical phenomena, such as size of mineral deposits, annual rainfall, and energies of meteorites and cosmic rays, follow Pareto distributions. The ready perception of structure in these phenomena is, as we saw in the density distribution of zero crossings of Peter and Paul's coin-tossing game, no guarantee that they are *not* due to pure chance. And, so Mandelbrot argues, the work required to validate the reality of any structure abstracted from a system displaying the statistics of Pareto exceeds by orders of magnitude the work expended hitherto on the validation of deterministic or first-stage indeterministic laws abstracted from systems displaying statistics where mean values rapidly converge to a limit. Thus until observational efforts of truly staggering dimensions can be expended on tests of second-stage indeterministic laws abstracted from systems for which there is no convergence to a limit of mean values at all, their scientific analysis will necessarily remain ambiguous. There does, of course, exist a spectrum of intermediate situations in which there occurs a *slow* convergence of mean values to a limit, and the effort required for analysis of such systems diminishes with the rapidity of this convergence.

Mandelbrot's main point, however, pertains to the future of the social sciences, particularly to economics and sociology. He signals, first of all, that the conspicuous lack of successful theories in these fields compared to the natural sciences cannot be ascribed (as is often done) to a difference in age. On the contrary, probability theory arose in connection with problems in the social sciences more than a century before indeterministic theories made their first appearance in physics. Hence indeterministic physics is younger than economics. No, the difference seems to arise from the predominance of Pareto distributions in the basic phenomena to which the social sciences must address their quantitative analysis. In economics, for instance, firm sizes, and income and price fluctuations follow Pareto's law. In sociology, the sizes of "human agglomerations" have a similar distribution, which demonstrates that such common-sense terms as "cities," "towns," and "villages" are ambiguous, impressionistic structures. That our vocabulary contains these terms, nevertheless, is a reflection of our habit of providing a specific description of a world whose events are intuited in terms of converging mean value statistics. Or, as Meyer has expressed it, "the redundancy which we are able to discover in the world is partly a function of the organization—the redundancy—built into the nervous system." Thus, according to this argument, it may be a vain hope to expect an early efflorescence of the social sciences, because most of their laws will be of the second-stage indeterministic kind. And hence verification of these laws would often require exertions that exceed by orders of magnitude all previous efforts expended on the natural sciences. Since it is not clear at present that such efforts are within the realm of the feasible, economics and sociology may long remain the ambiguous, impressionistic disciplines that they are at present. For it may be possible only in exceptional cases to ascertain whether their fundamental laws represent reality or figments of the imagination.

* * *

It seems to me that there is a strong formal resemblance between the nugatory semantics of transcendentalist art and the ambiguous epistemology of second-stage indeterminism in science. In both cases the percipient is more or less on his own, to make of his experience what he will. For all he knows, the events he witnesses are random in their origins. Thus arts and sciences, which in remote times both began as sublimations of the will to power and, meanwhile, have been traveling along separate paths, now appear to be approaching the same condition: Much work remains to be done, but how meaningful is it all?

Bibliography

Fiedler, L.A. *Waiting for the End*. New York: Delta Books, 1964.

Langer, Susanne K. *Philosophy in a New Key*. New York: Mentor Books, 1948.

Mandelbrot, B. New Methods in Statistical Economics. *Journal of Political Economy* 71, 421 (1963).

Meyer, L. B. *Music, the Arts and Ideas*. Univ. Chicago Press, 1967.

Price, D. J. de Solla. *Science Since Babylon*. New Haven: Yale Univ. Press, 1962.

Better Living Thru Chemistry in the Haight-Ashbury District of San Francisco, ca. 1965. Photographer: Edmund Shea. [By permission of Libra Artworks-American Newsrepeat Company, Berkeley.]

3

THE ROAD
TO
POLYNESIA

[1969]

After having outlined internal contradictions and limits of progress in the preceding two chapters, it is finally time to discuss the human condition which the putatively terminal stages of progress are now likely to bring about. As I indicated at the very outset of my exposition, I envisage this condition to be that of the Golden Age described by Hesiod more than twenty-five centuries ago. For the secular consequences of progress have now readied the Earth for that golden race of mortal men who, thanks to technology, will live like gods, without sorrow of heart, remote and free from toil and grief, but with legs and arms never failing, beyond the reach of all evil. In this chapter, I shall examine the coming of the Golden Age.

Before proceeding with this discussion, however, it is only fair to point out that, by the logic of my earlier argument, no scientific reliance can be placed on my projections for the future. For in the preceding discussion of the limits to the sciences, I adopted the view that "second-stage indeterminism" is likely to obtain in the analysis of social phenomena. And hence I am obliged to admit that the causal connections which I have previously inferred to exist between the events of the past and which I suppose to have given rise to the present cannot be presumed to lead to reliable predictions of coming events. My perceptions of structure in the historical record—the Bohemian phenomenon, the antitheses of will to power and economic security, the acceleration of progress, the trend toward freedom in artistic evolution, the exhaustion of scientific possibilities—are, for all I know, figments of my imagination that have no more reality than the gambler's perception of structure in Peter and Paul's coin-tossing record. And so one must necessarily consider my anticipation of the Golden Age an impressionistic vision rather than an objective forecast.

This, which appears on first sight to be a modest disclaimer on my part is, in fact, the height of presumption, and illustrates an important aspect of the present-day transcendentalist scene, namely the deflation of expertise. For my own dilettante psychologico-historicist analysis, based on a few months' reading of popular paperback books, is thus on a par with the best work of any professional who has devoted a lifetime of scholarship to these same matters. After all, according to this reasoning, the professional social scientist is no more able to demonstrate the validity of *his* inferences than I can of mine. It is, in fact, this deflation of expertise which made it possible that not long ago the canvas over which chimpanzee Betsy of the Baltimore Zoo had spread oil colors won a prize in a show of action painting.

I shall begin this chapter by summarizing a short book written about five years ago by the physicist Dennis Gabor, entitled *Inventing the Future*. This book has greatly influenced my own thinking, not so much by convincing me of the validity of its final prognosis, but by allowing me to see the problem of the future more clearly than I had seen it before. Undoubtedly, other writers have made more detailed and more professional analyses of the many topics discussed by Gabor, such as over-population, the future of capitalism, communism, and the underdeveloped countries, the limits of arts and sciences, and Common and Uncommon Man. But few of these writers have attempted the kind of global, mid-twentieth-century synthesis of all these facets which Gabor made in *Inventing the Future*.

Gabor starts out by positing what he calls the trilemma now facing mankind: nuclear war, overpopulation, and the Age of

Leisure. If either of the first two catastrophes is realized, mankind will be equipped to deal with it. The survivors of the holocaust world scramble back up to regain what was lost, and the hardiest among them would rebuild civilization. And the effects of overpopulation, life at the brink of starvation and confinement to narrow slave quarters, are only too familiar aspects from the past. But the third catastrophe, the advent of the Age of Leisure in which mechanization and automation will have rendered human labor largely superfluous, will find man's psyche unprepared, since leisure for all will be a complete novelty in human history. Boredom devolving from having no useful work to do might well lead mankind to a general nervous breakdown, similar to the psychic disturbances now not infrequent among the idle wives of the upper-middle class. Gabor writes, "in the past thirty years technology and social engineering have advanced with gigantic strides toward the Golden Age, with 'all the wonders that would be,' whose contemplation from afar was such bliss to the Victorian intellectuals, but very little has been done as yet to prepare us for it psychologically."

The gigantic strides toward the Golden Age were, of course, made only in the technologically advanced countries, while the majority of the world's population in the underdeveloped countries still lives in abject misery. But the advanced countries, particularly the United States and the Soviet Union and even, to an as yet more limited extent China, are already at work exporting their capital and technical know-how to the backward nations. And even though these exertions are not necessarily inspired by purely humanitarian motives, Gabor thinks it likely that through the inevitable ecumenical spread of technology the whole world will presently attain the same high standard of living. "Once industrialization has started," he says, "there is no stopping and no return." As far as the economics of this development are concerned, he reckons that even if the backward nations put none of their own incomes back into productive investments, an export of only 1 percent of the annual income of the "Free World" (or of 10 percent of its military expenditures) would suffice for the industrial takeoff of the backward nations. Gabor does not expect that this industrialization of the underdeveloped countries will occur within a democratic political framework, and he thinks that "if we try to impose unduly high democratic and moral standards on underdeveloped countries we shall not do them much good."

As far as the chances for nuclear war are concerned, Gabor finds some grounds for hoping that it can be avoided, in view of both the balance of terror and the manifest political rapprochement between the United States and the Soviet Union. He is, however, apprehensive of the possibility of China's becoming a nuclear power, which, were it to happen, "would be a black day indeed for China and the rest of the world." (Gabor's discussion did not foresee, of course, that in 1968, when that black day had come and gone, the neo-imperialist policies of both the United States and the Soviet Union still remained as greater threats to world peace.) As far as overpopulation is concerned, Gabor considers the population explosion in the underdeveloped countries a tragic but temporary phenomenon. Possibly millions of Asians will die of starvation before the end of the century—as adults instead of babies, as was the case formerly. But eventually, with increased industrialization and education, the birth rate will fall to adjust itself to the lower death rate. From the long-range point of view, it is more important to pay close attention to the population density of the advanced countries, for there it will be decided whether the equilibrium density of people is to be at the Malthusian starvation level or at a level more worthy of the dignity of man. In fact, Gabor thinks that in view of the modern means of transport the Western world is *already* overpopulated. And so he concludes that the archaic joy of having large families is the one luxury that civilization cannot afford. How to persuade young couples to avail themselves of birth control and forgo this joy is thus one of the most grave questions for the future.

Supposing then, hopefully, that nuclear holocaust can be avoided and that world population will stabilize at a tolerable level, one may inquire whether or not a long-term Age of Leisure is technologically feasible. In particular, it might be asked, will there not presently occur an exhaustion of the energy and mineral resources which man is presently squandering at an ever-accelerating rate? Gabor thinks there is good reason to expect that these problems can be successfully met. Admittedly, fossil fuels such as coal and oil will not be long for this world, but once, as Gabor expects it will, nuclear fusion power has become a going concern, our energy worries will be over for a very long time. But even if fusion power cannot be realized, then other, presently uneconomic but unlimited sources of energy, such as sunlight, tides, and volcanism will surely be exploited. The foreseeable exhaustion of high-grade metal ores is likely to be a more serious problem. But here Gabor expects that extraction of presently uneconomic but plentiful ores and the replacement of metal by plastics wherever possible will, in the end, meet this challenge successfully.

So now the age-old struggle against nature to vanquish poverty is nearly over. It has been a hard fight, won thanks to man's indomitable fighting spirit and the closing of the ranks between the knights of science and technology. But because of the ever-accelerating kinetics of progress, the state of economic plenitude arrived so suddenly that human nature has had no time to make the necessary adjustments. Gabor recalls that Moses, after showing his people the Promised Land, led them around in the wilderness for forty years, so that a new generation could grow up that would be worthy of it. According to Gabor, "the instinctive wisdom of the social body" has found the twentieth-century equivalent of the biblical wilderness, in which man can wander until the new generation is on the scene which is adapted to the Leisure Age. That wisdom is none other than "Parkinson's Law," which reduces the degree of leisure that our present technology could already afford by creating enough unnecessary work and waste. The ultimate psychological, as yet mainly subconscious, reason for the adoption of Parkinson's Law was epitomized by C. E. M. Joad: "Work is the only occupation yet invented which mankind has been able to endure in any but the smallest possible doses." (Freud, by the way, did not seem to share this opinion, since he took the down-to-earth view that the great majority of people work only under the stress of necessity and that it is the natural human aversion to work which raises most of the difficult social problems.) But now that the wasteful operation of Parkinson's Law *has* been widely recognized, Gabor believes, it cannot last for very much longer. The trek through the Parkinsonian wilderness will come to an end, and vast numbers of people, particularly those in the lower intelligence spectrum, will have nothing to do. By then the new generation had better be ready for the latter-day Promised Land, where the work of a very small and highly gifted minority, or Uncommon Man, keeps the majority in idle luxury. That majority, or Common Man, will be socially useless by the standards of our present-day civilization founded on the gospel of work.

Gabor now develops a series of eudaemonic propositions for meeting the threat of universal leisure. I shall not summarize them here because, in my opinion, they represent merely plans for a mid-twentieth-century intellectual's Utopia. To my mind, the major defect in these plans—education, eugenics, birth control, international solidarity—is that they ignore the motivational decay that is already in train. Gabor has by no means failed to note this trend; he makes such *aperçus* as that the ever-growing lack of hardship in the education of modern adolescents tends to make them less productive members of society, that the dedicated (and slightly mad) inventor is becoming a rarity, and that the ambitions of university students are not what they used to be. He does not, however, draw the lesson that these phenomena are but manifestations of the progressive loss of the will to power. But since the gospel of work is patently "the instinctive wisdom of a social body" that *has* the will to power, that gospel is bound to lose its charisma with the waning of the will.

In order to examine whether Joad's dictum that work is the only occupation yet invented which mankind has been able to endure in any but the smallest possible doses is really true, one must ask whether there have not, in fact, already existed affluent societies in recorded history in whose domain leisure was a prominent factor in everyday life. (Leisure *classes* that have lived on the backs of toiling masses in societies of general want are not, of course, what we want to consider here.) For if such affluent societies have existed, then their example should indicate to us how human nature can adjust itself to meet the problem posed by leisure. Gabor, despite his assertion that "leisure for all is a complete novelty in human history," is not unaware that instances of earthly paradises of leisure are, in fact, well-known. In this connection he mentions Burma, Bali, and the

South Sea islands "where people worked little and were satisfied with what they had." He describes also in some detail the happy and healthy Hunzas in their fertile Himalayan haunts—he duly notes that the Hunzas have no art—and finds that "it makes one gasp with surprise that human nature *can* be like this." But, for reasons I cannot fathom, Gabor believes that leisure afforded by a natural paradise and that by a technological paradise are entirely different matters. In contrast to Gabor, I believe that leisure is leisure and find it surprising, moreover, that the obvious relevance of the history of these paradises to our present condition is so rarely pointed out.

The history of the South Sea islands, or, more specifically, of Polynesia, can, I think, serve as a paradigm for the more general evolution toward the Golden Age. These islands were settled by a hardy and enterprising race, who set out some three thousand years ago eastward in open boats from Southeast Asia across the trackless emptiness of the Pacific in search of better homes. The voyages of these men represented daring feats of navigation in comparison to which the Mediterranean Sea traffic of the Phoenicians pales into insignificance. Even the much later sea voyages of the audacious Norsemen to Iceland, Greenland, and North America appear timid enterprises in comparison. As long as there still remained some Pacific terra firma to be discovered to the east and north, population pressure on the already settled territories caused adventurous splinter groups to venture farther into the unknown, carrying with them plants and animals for the stocking of virgin islands. By early Renaissance times, colonization of the Pacific was complete, and population control through infanticide and ceremonial cannibalism had been instituted. The colonists settled down to enjoy their exceptionally auspicious environment of abundant food, balmy clime, and relative rarity of natural enemies or adversities. Romanticized accounts have undoubtedly exaggerated the degree to which South Sea *vita* was *dolce*, but the general felicity of the environment does appear to have given rise to a typical personality not too different from the popular notion of the happy-go-lucky Polynesian. Though Polynesian society was by no means egalitarian, economic security for one and all was its dominant characteristic. Sensual gratification was a matter of primary

interest, while the not negligible dangers to the person presented by homicide and mayhem appear to have been faced with
surprising equanimity.

For the purpose of our present considerations, it is important
to note that at the time the European intruded upon this scene, a
very significant differentiation could be discerned in the direc-
tions and degrees to which Polynesian sociopsychological
evolution had progressed on the different islands. That is, the
more distant from the equator or the more barren and rugged
the territory, the greater the residual vigor, or what in present
American argot would be called the "straightness" of their in-
habitants. Possibly the most "straight" Polynesians were the
Maoris, whose ancestors had come to New Zealand in about A.D.
1000. These settlers populated a territory which was not only
much larger than any other of the islands settled by their race
but was also the only one so distant from the equator that it lies
squarely in the temperate zone. The Maori retained the enterprise of their ancestors, they were skilled agriculturists and arti-
sans, they possessed strong political organizations and formal
institutions of learning, and in their carving of wood and semi-
precious stones, maintained one of the few vital forms of
Polynesian art (the megalithic sculptures of the Marquesas and
Easter Island being another of the few instances of vital Polynesian art). The foremost factor in Maori life, however, was war,
which constituted its chief business and ideological mainspring.

On the other end of the social spectrum from New Zealand
were the Society Islands, in particular Tahiti. These islands,
where nature was at its most felicitous and vegetation at its
lushest, were settled at about the time of Christ. And here an
evolution set in which resulted in what we would now recognize as a beat society. In this hedonistic culture, neither religion
nor art, nor any kind of intellectual activity flourished. The Tan-
garoa monotheism of the enterprising settler-navigators had de-
generated into a formless pantheism, there was no laborious
sculpture of colossal stone statuary, and the art of pottery and
the use of ideographic writing were lost. And precisely that
aspect of Polynesia, and of Tahiti in particular, which has inspired so much of the romanticization since its discovery by
Europeans is also of interest for us here because of the obvious

analogous evolution in our affluent society: its sexual mores. Evidently the repression of the sexual drive, a nearly ubiquitous and supposedly very ancient aspect of human nature, suffered an extensive derepression in the paradise of the South Seas. Sexual promiscuity among adolescents was the general rule, and though the custom of marriage among adults was still retained, the structure of the resulting family became very loose. Serial polygamy—easy and frequent divorces and remarriages—obtained, and though adultery remained formally proscribed, its occurrence was very common. The sexual license of Tahiti found its apotheosis in the Arioi Society. This society, which appears to have arisen as a magico-religious sect in earlier days, developed into an organization of traveling performers of what by European standards were highly obscene rites. The male and female *sociétaires* possessed each other in common, and society rules demanded that all offspring resulting from their unions were to be killed at birth. Another aspect of Polynesian life highly relevant to our affluent society is the important role played by kava, a psychedelic drug extracted from the root of the plant *Piper methysticum*. In its use of kava, as in its sexual practices, Tahiti seems to have shown an extreme development. Whereas at the time of the first European visits kava drinking was confined mainly to highly ritualized ceremonial occasions in western Polynesia, in Tahiti kava was in free use for frequent, personal hallucinatory trips.

Quite apart from any restrictions imposed on historical interpretation by "second-stage indeterminism," it is in any case obviously dangerous to prophesy the future on the basis of historical precedent. However similar some earlier situation might appear to the present, one seemingly trivial difference between then and now, there and here, might, in fact, be of such great importance for our destiny that it could easily vitiate the predictive value of any comparison. And thus one must be careful not to overstrain the analogy between Polynesia and the coming Golden Age, in which technology will soon provide for Everyman what a felicitous constellation of natural circumstances once provided for the South Sea islanders. But, if nothing else, the history of Polynesia does show that the "threat" of leisure was met at least once before by simply and easily abandoning the gospel of work. It shows that people will not necessarily go stark, raving mad when, in a background of economic security, most of them no longer have much useful employment. Furthermore, that history lends additional support to the notion I tried to develop earlier that economic insecurity is a necessary condition for the paragenetic transmission of the will to power, and *a fortiori* for the perpetuation of the pinnacle of its sublimation: Faustian Man. The Vikings of the Pacific must have started out on their eastward trek with a strong Faustian bent, but by the time Captain Cook found them, Faustian Man had all but disappeared from the Society Islands.

*   *   *

The Polynesian example now allows us to perceive that even though, as Gabor says, in the past thirty years technology and social engineering have advanced with gigantic strides toward the Golden Age, it is *not* true that very little has yet been done to prepare us for it psychologically. On the contrary, the rise of beat philosophy engendered by negative feedback of the accession to economic security in the affluent society on the will to power is precisely such a preparation. Obviously, the prospect of universal leisure holds little terror for beat society.

But though the beat, or Polynesian answer to the leisure problem posed by the Golden Age is certainly one feasible solution, it might not necessarily follow that it is the only one. Gabor, for instance, puts his hope in the appearance of what he calls "Mozartian Man," a hypothetical creative type of which he believes Mozart to be a premature forerunner, "whose art does not live on conflict but who creates for joy, out of joy." Mozartian Man will make the most of the creative opportunities provided by the leisure afforded him and thus furnish inspiration for his less gifted and mainly unemployed fellow men, who, through appropriate education, he keeps from drink and crime. A similar optimistic view of the Golden Age has been developed by another physicist, John Platt, in his book *The Step to Man*. Unlike Gabor, who *is* worried but hopeful, Platt is positively elated by the prospect of universal leisure. He envisages that man will at last be liberated from the shackles of menial toil and can now proceed to consecrate his boundless energy, hitherto largely wasted on base activities, entirely to higher, creative things. But as I have tried to show, the "creative" activity of Gabor's Mozartian Man, or of Platt's equivalent, will, in any case, be of a qualitatively different nature from what, I daresay, these authors had in mind. In the arts, our future Mozart is most unlikely to bear any resemblance to his namesake. He will be either a transcendentalist whose creations are not intended to convey any meaning, particularly not joy, or an epigone working in one of the traditionally semantic styles which had seen its heyday long ago. In the sciences, our future genius will be similarly engaged in activities whose significance would be unlikely to make a deep impression on Gabor or Platt. He might be working out the detailed genetic map of yet one more species of bacteria or searching for one more class of subatomic particle. Or he might be a social scientist who is developing yet one more subjective interpretation of data whose statistical nature puts them beyond the pale of successful theoretical formulation. He might even be collecting rock samples on Mars, in which case we could ask him, as Arthur Koestler wanted to ask Space Cadet Tom Corbett (in a quotation cited by Gabor): "Was your journey really necessary?"

It must be obvious that beat philosophy and transcendentalism are closely affined, even though I had previously discussed them within somewhat different contexts. Evidently, the inner-directed, antirational attitude of the beatnik renders him the natural audience for the meaningless works of transcendentalist art. Beat philosophy, furthermore, seems to be exactly the suitable psychic infrastructure for the scientist of the future, in view of the previous inference that the sciences are rapidly approaching the limits of their significant progress. The beat scientist would derive his satisfaction from the experience of merely being in his laboratory and doing experiments that are meaningful to *him*. Whether the results he obtains are really original, correct, or of significance for anyone else is of small concern. In this way, science can go on and on, though, like art, it will bear only a superficial resemblance to what had been understood by that term in the past. Indeed, in addition to meeting the threat of universal leisure, the rise of beat philosophy will get mankind off another horn of Gabor's trilemma: nuclear war. As far as I can see, a beat society provides a much more durable guarantee against atomic holocaust than the balance of terror. Presently no one will be any longer *interested* in such outer-directed expressions of the will to power as the making of war. In any case, ideology and economics, traditionally the two principal causes for war, will have lost most of their relevance for the transcendentalists of the Golden Age.

Finally, I want to consider the most recent Bohemian phenomenon, namely the hippies, whose appearance in the Haight-Ashbury district of San Francisco in 1966 signaled a further adaptive step toward the Golden Age. I first became aware of this fact when I saw Lucas Cranach's painting "The Golden Age" on a visit to the Munich Pinakothek Museum. It suddenly dawned on me that the subject of Cranach's four-hundred-year-old painting was nothing other than a prophetic vision of a hippie be-in in San Francisco's Golden Gate Park. The hippies are evidently a successor phenomenon to the beatniks, from whom they have taken over the inner-directed, antirational, existential attitude. However, the hippies have thrown overboard some more pieces of the traditional motivational baggage and retain but the tiniest residue of the once mighty will to power. With the appearance of the hippies there has become manifest a metamorphosis of the traditional human psyche even more radical than the resignation of the will to power, namely an erosion of what Freud called the *reality principle*. According to Freud, for some time after birth the child's self includes the sum total of its experiences, of both internal and external provenance. Only at later stages of its development does the child come to distinguish between these two experiential sources. It restricts the self to the world of internal events and begins to construct an external reality on the basis of the emanations from the world of outside events. The capacity to make this distinction is, it goes without saying, of the utmost survival value, and its failure is, according to Freud, the cause of important psychotic syndromes. Needless to say, possession of the reality principle is a precondition for possession of the will to power, through which the self seeks hegemony over events in the outside world. Freud outlined several, not necessarily mutually exclusive, avenues toward a deflation of the reality principle. One of these is represented by a reduced shrinkage of the infantile, all-inclusive self. In such persons, the adult self still encompasses many of the events of the outside world, a condition which Freud referred to as an "oceanic feeling," a feeling of oneness with the universe. Another way of deflating the reality principle is the willful reduction of the import of outside events, either by the use of drugs or by the control of the instincts, as in the practice of Yoga. It is of no little significance that both of these assaults on the reality principle form important parts of the teaching of Oriental philosophies that are now finding ever-greater resonance in the West. In some sense, the reality principle suffered premature erosion in the East. For the economic productivity of those societies had reached a level high enough for only a tiny fraction of the people to embrace views so clearly detrimental for survival in a *de facto* hostile nature. Indeed, it may well be that widespread, *partial* adoption of these philosophies to the degree still compatible with physical survival in such countries as India and China led to the later stagnation of these formerly dynamic civilizations. But in the leisure society of the Golden Age, adhesion to the reality principle will no longer be so critical for survival.

Although the beatniks had greatly deflated their will to power and thus resigned much of the ambition to change the outer world, they nevertheless still seemed to maintain considerable contact with reality. Thus the sense experiences from which the inner-directed realization of the self was sought were in the main of external provenance, as testified to by the interest of the beatniks in such activities as travel, food and drink, jazz, poetry, and sex. The use of hallucinatory drugs, though not foreign to the beatniks, did not assume nearly the importance that it was to reach on the hippie scene a decade later. Now, however, the much more extensive use of drugs as an experiential source has brought about a rather far-reaching abnegation of reality, or as its prophet of psychedelia, Timothy Leary, calls it, "dropping out." That is, the boundary between the real and the imagined has been dissolved. For the hippies, the reality principle is all but dead. This overt erosion of the reality principle embodied in the hippies was not, of course, invented in the Haight-Ashbury district. On the contrary, the philosophical basis of reality has been the subject of critical discussions for some two hundred years, ever since Immanuel Kant claimed that things-in-themselves are unknowable and that our notion of reality is a construct of human reason. The transcendentalist world picture of the present avant garde artists, mentioned in the preceding chapter, is evidently another latter-day reflection of this trend to lessen the importance of distinguishing between the real and the imagined. The lessening of this distinction appears also to be the theme of such latter-day films as Resnais' *Last Year at Marienbad* and Antonioni's *Blow-Up*. But the novelty of the hippies consists in their being the first large-scale community in the West which actually *acts* according to these ideas.

Finally, I will attempt to synthesize my preceding considerations into an image of the coming Golden Age. This synthesis must obviously assume that there will be no nuclear war, an assumption which is based mainly on optimism. But failure of that assumption would render all present considerations of man's future nugatory in any case. If nuclear war can be avoided in the *near* future, then I believe that the general waning of the will to power will presently lead to a condition in which the holocaust will become increasingly less probable, because interest in war will be largely gone. Following Gabor's projections, I believe that the presently underdeveloped nations will, sooner or later, reach the same level of economic affluence as that presently enjoyed by the technologically advanced nations. These economic changes will, in turn, engender the global hegemony of beat attitudes, which, in the Orient, at least, are already deeply rooted in philosophical tradition. I also assume that there will be no technological or biological developments as radical as the achievement of travel faster than the speed of light or the enlargement and structural alteration of the human brain. Failure of this latter assumption would set off an entirely new phase in human evolution, whose course cannot be envisaged by simple extension of past history.

With these assumptions one arrives at the conclusion that the Golden Age will not be very different from a re-creation of Polynesia on a global scale. (It is not unreasonable to expect that the high rate of infanticide and homicide of old-time Polynesia will not be a feature of the Golden Age, since more humane means of avoiding overpopulation are now available.) Though there will never be enough Tahitis to accommodate the world's population, comfortably air-conditioned metropolitan apartments will easily provide a satisfactory ersatz for authentic beachcombing. The will to power will not have vanished entirely, but the distribution of its intensity among individuals will have been drastically altered. At one end of this distribution will be a minority of the people whose work will keep intact the technology that sustains the multitude at a high standard of living. In the middle of the distribution will be found a type, largely unemployed, for whom the distinction between the real and the illusory will still be meaningful and whose prototype is the beatnik. He will retain an interest in the world and seek satisfaction from sensual pleasures. At the other end of the spectrum will be a type largely unemployable, for whom the boundary of the real and the imagined will have been largely dissolved, at least to the extent compatible with his physical survival. His prototype is the hippie. His interest in the world will be rather small, and he will derive his satisfaction mainly from drugs or, once this has become technologically practicable, from direct electrical inputs into his nervous system. This spectral distribution, it will be noted, bears some considerable resemblance to the Alphas, Betas, and Gammas in Aldous Huxley's *Brave New World*. However, unlike Huxley, I do not envisage this distribution to be the result of any purposive or planned breeding program, but merely a natural population heterogeneity engendered mainly by differences in childhood history. Furthermore, in contrast to the low-grade producer roles assigned to Betas and Gammas, beatniks and hippies will play no socioeconomic role other than being consumers.

As far as culture is concerned, the Golden Age will be a period of general stasis, not unlike that envisaged by Meyer for the arts. Progress will have greatly decelerated, even though activities formally analogous to the arts and sciences will continue. It is obvious that Faustian Man of the Iron Age would view with some considerable distaste this prospect of his affluent successors, devoting their abundance of leisure time to sensual pleasures, or what is even more repugnant to him, deriving private synthetic happiness from hallucinatory drugs. But Faustian Man had better face up to the fact that it is precisely *this* Golden Age which is the natural fruit of all his frantic efforts, and that it does no good now to wish it otherwise. Millennia of doing arts and sciences will finally transform the tragicomedy of life into a happening.

Bibliography

Buck, P. H. *Vikings of the Pacific.* Univ. Chicago Press, 1959.

Gabor, D. *Inventing the Future.* Harmondsworth, England: Penguin Books, 1964.

Heilbronner, R. L. *The Future as History.* New York: Grove Press, 1961.

Huxley, A. *Brave New World Revisited.* New York: Perennial Library, 1965.

Platt, J. R. *The Step to Man.* New York: Wiley, 1966.

Suggs, R. C. *The Island Civilizations of Polynesia.* New York: Mentor Books, 1960.

Williamson, R. W., and R. Piddington. *Essays in Polynesian Ethnology.* Cambridge Univ. Press, 1939.

II

MOLECULAR GENETICS IN THE SALON

Francis Crick and James Watson, sliding down their DNA Double Helix. [From the cover of the special issue of Nature *(vol. 248, no. 5451, 1974) commemorating the 21st anniversary of the publication of the letters announcing the discovery of the structure of DNA. Reproduced by permission of Macmillan Journals Ltd.]*

4

WHAT
THEY ARE
SAYING
ABOUT
HONEST JIM

[1968]

Just as the Greeks divided the history of man into Golden, Silver, Brass, Heroic and Iron Ages (but erring in having history starting from, rather than leading up to, the Golden Age), so can we divide the history of molecular genetics into a succession of distinct periods. The first of these, or *Classic Period* (corresponding to the Iron and Heroic Ages), began in remote, neolithic antiquity, at the dawn of civilization with its first moves towards breeding domestic plants and animals. It lasted until the 1940's, when the function of genes, the discrete hereditary factors discovered by Gregor Mendel in mid-nineteenth century, was recognized to consist of the governance of formation of specific enzymes. The Classic Period gave rise to a corpus of knowledge that accounted successfully for the role of genes in heredity and evolution, but in formal rather than molecular terms. The second, or *Romantic Period* (corresponding to the Brass Age), began in the 1940's, when a small, tightly knit group of investigators, many of them trained in the physical sciences and having Max Delbrück at their ideological focus, took an interest in genetics, in the hope of explaining the mystery of biological self-reproduction in molecular terms. The Romantic Period defined clearly the problems that molecular genetics was to solve eventually, and it set the tone and working style of the subsequent periods. The Romantic Period ended in 1952, when general recognition came to Oswald Avery's earlier discovery that the hereditary material—the substance of the genes—consists of DNA. The third, or *Dogmatic Period* (corresponding to the Silver Age), began in 1953, when James Watson and Francis Crick discovered the double-helical structure of DNA. In the wake of this discovery, molecular solutions for the deep problem of how the DNA manages to carry out its self-replication and governance of formation of specific enzymes were put forward, as a network of dogmas whose truth seemed self-evident. The Dogmatic Period ended in 1961, when experimental proof of the general validity of this dogmatic weft had been delivered, and when it had become possible to break the genetic code that specifies the semantic relation between the hereditary information stored in the DNA and the structure of the enzymes, which represents the meaning of that information. The fourth and final, or *Academic Period* (corresponding to the Golden Age), was solemnly inaugurated in 1962 by Gustaf VI of Sweden, when he awarded the Nobel Prize to Watson and Crick. This period, which still obtains today, has seen fantastic technical achievements, foremost among them the total decipherment of the genetic code and a vast extension of our molecular understanding of the hereditary material. It has also brought the first practical fruits of molecular genetic knowledge to medicine and public health. As a subject matter for future research, molecular genetics is far from exhausted. But its appeal as an arena for Faustian strife against the mysterious unknown is mainly gone.

As molecular genetics was settling comfortably into its Academic Period in the late 1960's, molecular geneticists began to find enough time to write non-technical books meant to acquaint the lay public with the scientific achievements, historical origins, and philosophical implications of their discipline. In other words, efforts now got underway to extend the triumphs of molecular genetics from the laboratory to the salon. The collection of autobiographical essays *Phage and the Origins of Molecu-*

lar Biology (J. Cairns, G. S. Stent, and J. D. Watson, eds., Cold Spring Harbor, 1966), brought out in honor of Max Delbrück's sixtieth birthday by his fellow veterans of the Romantic Period, was one of the first of these undertakings. Though this collection became a *succès d'estime*, well-received by a limited professional readership, it failed to reach a wider audience. But a very different fate was to meet *The Double Helix* (Atheneum, Boston, 1968), James Watson's account of his and Crick's discovery of the structure of DNA. Publication of personal reminiscences by famous scientists of the work that made them famous are commonplace enough, but there was evidently something highly unusual about Watson's book. In contrast to the run-of-the-mill autobiographical writings in this genre, which are, as a rule, of interest only to experts already familiar with both the scientific achievements described and the names associated with them, *The Double Helix* became an international bestseller, enjoyed by readers with little prior knowledge of molecular genetics and quite unaware of the existence of Watson, Crick, or any of the other persons appearing in this story. Moreover, Watson's book raised a storm of controversy among biologists, for whom it provided the major off-duty conversational topic in the months following its appearance.

The first intimation that there might be something special about Watson's book was given by prepublication newspaper stories, which revealed that in the previous year Harvard University Press had been ordered by its suzerain, the Harvard Corporation, to renounce its agreement to publish *The Double Helix*. These stories caused Watson's new publisher, Atheneum Press, a commercial house not unaware of the sales appeal of Greater-Boston-banned literature, to increase greatly its initial print order of the book, a decision which, in view of later sales records, was most wise. *The Double Helix* was widely reviewed, and the ensemble of reviews turned out to provide some interesting insights into the notions that are held in diverse quarters regarding the work and lifestyle of scientists. Thus, when asked to review Watson's book myself, after a flood of reviews had already appeared, I decided to engage in the kind of derivative, second-order scholarship which in British literary circles is often alleged to be typical of American academics, namely, studying not the authors but their critics. As I intend to show here by a brief analysis of six selected reviews of *The Double Helix*, the relativity principle illustrated by the movie "Rashomon" was at work: although all six reviewers are obviously writing about a book by one J. D. Watson, describing the discovery of the structure of DNA and presenting characters named Francis Crick, Maurice Wilkins, Linus Pauling and Rosalind Franklin, one would hardly think that all six reviewers actually read the same book.

Undoubtedly the most important review, in terms of its widespread circulation, was that printed over Philip Morrison's name in *Life Magazine* (March 1, 1968). I find it difficult to believe that someone of the stature of Morrison, a Professor of Physics at M.I.T., actually wrote this piece. (How, for instance, could Morrison have written that passage of his review which says that in Watson's book "there is plenty of clearly put talk about atoms, molecules and hydrogen *bombs*" (italics mine)?) Though the *Life* review points out correctly that this book is an autobiographical account of how Watson "discovered how DNA molecules look" and that "the idea for the double helix had to come somehow, in the way of great new ideas," it has not much else to recommend it in the way of accuracy or depth. Thus *Life* readers, opening their copy of *The Double Helix* expecting to find "the air of a racy novel of one more young man seeking room at the top," would be justified in their belief that they had been had, for Watson certainly does not deliver six dollars' worth of "censored movies, smoked salmon, French girls, tailored blazers [which] set the stage on which ambition, deft intrigue and momentary cruelty play their roles." Though Watson is said to have "a sharp eye and an honest tongue," the review evokes a false impression of gentility and niceness, qualities whose absence from *The Double Helix* forms one of its most striking features. Thus, in Copenhagen Watson is said to have "learned chemistry from an 'obviously cultivated' man," whereas Watson actually states that the plan to learn DNA chemistry in Copenhagen was "a complete flop," and that the only time Watson understood what that man was saying had been when he announced that his marriage was over. And Rosalind Franklin is reported to be "rightly and warmly praised for her key x-ray work," whereas one of the chief points of Watson's story is, of course, that Franklin's stubborness was a major obstacle for Wilkins' working out the structure of DNA; the posthumous praise of Franklin's attainments is relegated to an obviously grafted-on epilogue. The review closes with a reply to the question: "How's Honest Jim?." Apparently blind to the central significance for the whole book of the episode in which Watson is sarcastically greeted in this manner, *Life* inanely lets "one reader answer for him: Fine, just fine."

Ascending the ladder of sophistication to the next rung, we may consider the review in the *Saturday Review* (March 16, 1968) by John Lear, the science editor of that serial. Lear begins his review by saying that "this book is being acclaimed as the Pepys diary of modern science." He then proceeds to prove that this acclaim is unjustified because Watson, unlike Samuel Pepys, was not the Secretary of the British Admiralty and neither participated in the restoration of Charles II nor endured the visitation of London by the plague. Also, Watson's writing style, according to Lear, has little distinction. So how can "The Double Helix" resemble Pepys' diary? Good points, those! Lear does not bother to state, however, *who* has actually made the claim he troubles to demolish. Surely not Sir Lawrence Bragg, who, in his foreword to *The Double Helix*, merely suggests that Watson "writes with a Pepys' like frankness." Indeed, it seems to be precisely Watson's frankness that caused Lear to wonder "what qualities of achievement [Watson's] Nobel award was intended to celebrate." Was it, as "Watson reveals with no apparent regret, his hope that his pretty sister would serve as a romantic decoy in obtaining otherwise inaccessible information essential to his research . . ." or his use of "his young friend Peter Pauling to spy on Pauling's brilliant father Linus . . ." or "his attempt to bully a proud woman scientist into discussing details of her X-ray studies of DNA"? If it really *was* the intention of the Nobel Committee to celebrate these particular qualities, then it made a serious mistake, for Watson wrote that he merely hoped that Maurice Wilkins' interest in Elizabeth Watson might allow him to join Wilkins' research group; that Peter merely told him that Linus had worked out a structure of DNA and later showed him the by no means secret manuscript describing this structure; and (in a lengthy passage quoted by Lear) that he was merely trying to escape from the laboratory of Rosalind Franklin, who, he feared, was about to strike him. Surely instances of more substantial villainy could have been made known to the Committee.

Lear is worried that *The Double Helix* may have a corrupting effect on the impressionable minds of high school and college students, who, in their idealism, may turn away from becoming scientists once they learn how Watson gained his Nobel prize by knavery. "Fortunately for the future of science, they will acquire a certain amount of perspective from the knowledge that the two men who got the 1962 prize with Watson objected to the text of *The Double Helix* with sufficient vigor to encourage the university press of his home campus—Harvard—to abandon the book's publication." Though Lear's idea of inspiring idealistic youths through preventing the publication of books may not be mainstream *Saturday Review* thought it certainly *is* familiar to John Birch Society adherents. Lear tries to tell the history of genetics, in order to show that Watson's forerunners such as Darwin, Mendel, Miescher and Morgan were all modest fellows who, unlike Watson, did not claw their way to public attention. Lear does not have his facts quite straight. He falsely asserts that "it was the great Charles Darwin who first aroused wide interest in inheritance by promulgating the theory of evolution." In fact, such "wide interest" came only at the turn of this century with the rediscovery of Mendel's work, for which rediscovery the study of development rather than evolution paved the way. Equally groundless is Lear's statement that Darwin did not learn of Mendel's laws of inheritance because Mendel "had as little interest as Darwin did in personal aggrandizement." For Mendel is known to have sent out reprints of his papers to several leading biologists of his day, who simply did not grasp the significance of his work. And contrary to Lear's assertion, it was not the "Darwinian sense of fair play [that] required simultaneous publication with Wallace" but the Darwinian fear of getting scooped. Finally, Lear attributes chemical mutagenesis to Morgan, which was, in fact, discovered by Auerbach and Robson when Morgan was 75 years old. In any case, even if Lear's account *were* accurate, all it would prove is that Watson's predecessors did not write their "Double Helix," no more than did Watson write Pepys' diary. In a final twisting of his blunt knife, Lear suggests that Watson's contribution to the discovery of the DNA structure was not all that great anyway. It was obvious, he intimates, that "there would be in the DNA molecule a spiral stairway with steps in a particular order. The question that remained to be decided was whether the step-plates were within or outside the spiral." This, of course, is a factious distortion of the ideological situation facing Watson and Crick at the outset of their work. At that time the idea that DNA encodes genetic information in the form of a particular nucleotide (or step-plate) order was virtually unknown and it was even less apparent that this order is embodied in a helical, let alone *double* helical, molecule. I wonder what effect Lear's review will have on impressionable minds considering book-reviewing as their life's work. His gall will surely turn off the critical aspirations of any idealistic kid.

We now mount several more rungs on our ladder to reach the next two reviews. The first of these appeared in *The Nation* (March 18, 1968) and was written by J. Bronowski, Senior Fellow in the Salk Institute for Biological Studies and veteran author on matters concerning the social implications of science. Bronowski evidently saw the privately circulated earlier draft of *The Double Helix*, because he notes with some regret that the published version has been "bowdlerized here and there" as a result of objections raised by some of the main protagonists of Watson's story. Nevertheless, Bronowski finds, though some of "the small darts of fun and barbs of malice" are gone, the book has not lost its savor. It still remains "a classical fable about the charmed seventh sons, the anti-heroes of folklore who stumble from one comic mishap to the next until inevitably they fall in to the funniest adventure of all: they guess the magic riddle correctly. Though the traditional parts of Rosalind Franklin as the witch and Linus Pauling as the rival suitor have been toned down . . . , they are still unmistakably what they were, mythological postures rather than characters." Bronowski finds that Watson has managed to tell that fairy tale with the quality of innocence and absurdity that children have. "The style is shy and sly, bumbling and irreverent, artless and good-humored and mischievous. . . ." But maybe Watson is not all that artless after all, since Bronowski also recognizes him as playing Boswell to Crick's Dr. Johnson—"monumentally admired, and (every so often) scored off." (Fortunately, it seems Lear had not gotten wind of the Watson-Boswell analogy).

Bronowski finds that the importance of Watson's book transcends the mere telling of a good story, however, in that "it communicates the spirit of science as no formal account has ever done. . . . It will bring home to the nonscientist how the scientific method really works: that we *invent* a model and then test its consequences, and that it is this conjunction of imagination and realism that constitutes the inductive method." Another important general point brought out by the book is the importance of ruthless criticism for the progress of science: ". . . if you cannot make it and take it without anger, . . . then you are out of place in the world of change that science creates and inhabits."

So far, so good. But in the closing paragraph of his review Bronowski expands his considerations to the general literary scene, and now reveals a view of the relation of contemporary writing to the actual human condition which I find surprising for a man with Bronowski's interests and experience. In that paragraph he seems to abandon his earlier finding that *The Double Helix* is a classical fable in the folkloristic tradition and asserts that "its two happy, bustling, comic anti-heroes are new in literature today, and yet they should be a model for it, because they run head-on against the nostalgia for defeat which haunts the writer's imagery of action now." Bronowski does "not suppose *The Double Helix* will outsell Truman Capote's *In Cold Blood* but [thinks that] it is a more characteristic criticism and chronicle of our age, and [that] young men will be fired by it when Perry Smith and Dick Hickock no longer interest even an analyst." How could Bronowski have failed to realize the obvious parallelism between the two books he posits here as antitheses? Quite apart from the strictures of bad taste made by many critics against both Watson and Capote for writing their "true novels," Capote's Smith-Hickock anti hero pair shares with Watson's self-portrait one essential feature of capital analytical interest: the finding of only transitory existential meaning in action.

The other review which we reach at this level is one in *Science* (March 29, 1968) by Erwin Chargaff, Professor of Biochemistry at Columbia University. Chargaff, as discoverer of the compositional adenine-thymine and guanine-cytosine equivalence in DNA, has an important part in the story told by Watson. To some readers, unfamiliar with Chargaff's speeches and writings over the past dozen years, his review must have seemed surprisingly sarcastic; to other readers, aware of Chargaff's long-standing lack of appreciation of the achievements of Watson and Crick in particular and of the working style of molecular biology in general, the review may have seemed unexpectedly mild. At the very outset Chargaff plays one of his old gambits: stating, *en passant*, that Watson and Crick "popularized" purine-pyrimidine base pairing in DNA. Readers familiar with the Chargaff auto-anthology, "Essays on Nucleic Acids," will understand that this parlance is to imply that he, Chargaff, and not Watson and Crick, really discovered base-pairing. But if Chargaff *did* discover base-pairing before Watson and Crick, then not only did he not "popularize" it, but he kept it to himself until well after it had become popular.

Chargaff finds that though Watson is not as good a writer of garrulous prose as Sterne, he *has* managed to pull off a "sort of molecular Cholly Knickerbocker" (Lear was only willing to rank Watson with Walter Winchell). Like Lear, Chargaff is bothered by the Pepys' diary analogy but, unlike Lear, he does quote Bragg's foreword. What is significant for Chargaff in this connection is that Pepys, unlike Watson, did not publish his frank observations during his lifetime. Chargaff seems to imply, without actually saying so overtly, that publishing frank impressions of one's contemporaries is rather poor taste, though he does admit that he is not above enjoying some of Watson's revelations about Crick.

Chargaff declares, with some justice, that Watson's book belongs to the realm of scientific autobiography, a most awkward literary genre. Most such books, he says, give "the impression of having been written for the remainder tables of bookstores, reaching them almost before they are published." The reasons for this, according to him, are not far to seek: scientists "lead monotonous and uneventful lives and . . . , besides, often do not know how to write." *The Double Helix*, as Chargaff admits, is certainly a quite exceptional member of this genre and may not make the remainder table scene for quite some time. Chargaff now attempts to provide a more general reason for the triteness of scientific autobiographies, namely that whereas *Timon of Athens* could not have been written and "Les Desmoiselles d'Avignon" could not have been painted had Shakespeare and Picasso not existed, in science the rule is that "what A does today, B or C or D could surely do tomorrow." Quite apart from the intrinsic impossibility of subjecting this view of artistic and scientific evolution to any test, and hence quite apart from its nugatory historicism, Chargaff surely realizes that the foundation of great writing is depth rather than uniqueness of experience. So why does he bother to make this empty point? Aha! Watson and Crick are not such great fellows anyhow, because, if they had not found the structure of DNA in March of 1953, someone else would have surely found it by the following April.

Chargaff, like Bronowski, closes his review with a wistful look at the Good Old Days, compared to which Things Have Now Gone to the Dogs. But what a difference in their views of past and present! Bronowski, on the one hand, sees Watson and Crick as chips off the old block, ambitious, hard working, adventurous, and optimistic, in felicitous contrast to the defeatist New Generation. Chargaff, on the other hand, considers them typical of the "new kind of scientist," who "could hardly have been thought of before science became a mass occupation, subject to, and forming part of, all the vulgarities of the communications media." Chargaff does not spell out his ideas of the "old kind of scientist," though I suspect that what he has in mind is Paul Muni playing the part of "Louis Pasteur."

Another step on our ladder brings us to what I would consider one of the two substantial reviews among the six I am considering here, namely that published in *The New York Times Book Review* (Feburary 25, 1968) by Robert K. Merton, Professor of Sociology at Columbia University. First of all, Merton finds that this is not just one more scientific autobiography, in that Watson is describing the events that led up to one of the great biological discoveries of our time. This finding is thus in stark contrast to that of Merton's colleague Chargaff, who views Watson mainly as a successful popularizer of notions that were already in the air. Merton says that he knows of nothing quite like it in all of the literature about scientists at work. Furthermore, since Watson is "telling it like it was," or at least as it seemed to the then youthful Jim, the book is an important contribution to scientific historiography. "The public record of science tends to produce a mythical imagery of scientific work, in which disembodied intellects move toward discovery by inexorably logical steps, actuated all the while only by the aim to advance knowledge." Watson sets this record straight, in showing "a variety and confusion of motives, in which the objective of finding the structure of DNA is intertwined with the tormenting pleasures of competition, contest and reward. Absorption in the scientific problem alternated with periodic idleness, escape, play and girl-watching. Friendship and hostility between collaborators [were] expressed in a nagging yet productive symbiosis in which neither could really do without the special abilities of the other. And all this engaged not only the passion for creating new knowledge but also the passion for recognition by scientific peers and competition for place."

Merton understands rather more about the sociology and history of science than do Lear and Chargaff, for he signals that competition and property rights in science are as old as modern science itself. (By "modern" I imagine Merton means "post-Renaissance" and not the latter-day period of Chargaff's "new kind of scientist"). The novelty of Watson's story is merely that he has so revealingly described this element for the general reader. For it is important to realize that the operation of the scientific community cannot be understood from the premise that the advancement of knowledge is its only institutionalized motive. Why, Merton asks, is science so competitive? Is it because it "tends to recruit egotistic personalities, contentious and exceedingly hungry for fame?" No, "the competitive behavior of scientists results largely from values central to the scientific enterprise itself. The institution of science puts an abiding emphasis on significant originality as an ultimate value, and demonstrated originality generally means coming upon the idea or finding first. Recognition and fame thus appear to be more than merely personal ambitions. They are institutionalized symbol and reward for having done one's job as a scientist superlatively well."

Finally, after taking a few more upward steps we reach the second of the two substantial reviews and attain the pinnacle of our ascent. For we now behold the most excellent article in the *New York Review of Books* (March 28, 1968) written by Sir Peter Medawar. Medawar begins his review by explaining that the significance of the discovery by Watson and Crick went far beyond "merely spelling out the spatial design of a complicated and important molecule. It explained how that molecule could serve genetic purposes. . . . The great thing about their discovery was its completeness, its air of finality. If Watson and Crick had been seen groping toward an answer, if they had published a partly right solution and had been obliged to follow it up with corrections and glosses, some of them made by other people; if the solution had come out piecemeal instead of in a blaze of understanding: then it would still have been a great episode in biological history, but something more in the common run of things; something splendidly well done, but not done in the grand romantic manner." Medawar also points out that in the years following their discovery of the DNA double helix, Watson and Crick showed the way towards the analysis of the genetic code and the understanding of how the genetic material directs the synthesis of proteins. He finds that "it is simply not worth arguing with anyone so obtuse as not to realize that this complex of discoveries is the greatest achievement of science in the twentieth century."

As far as the sense of keen competition conveyed by Watson's story, and the possible shock experienced by lay readers over the revelation that science is not a disinterested search for truth, Medawar is one with Merton in declaring that the notion of indifference to matters of priority is simply humbug. For what accomplishment, he asks, can a scientist call "his" except those things that he has done or thought of first? This does not mean, however, that meanness, secretiveness and sharp practice are not as much despised by scientists as by other decent people in the world of ordinary everyday affairs. Medawar finds, however, that for a person as priority conscious by his own account as Watson, he is not very generous to his predecessors. Why, in particular, did he not give a little more credit to people like Fred Griffith and Oswald Avery, whose work on bacterial transformation had demonstrated that DNA is the genetic material? Medawar's explanation is that this happened not for a lack of generosity but for simply being bored stiff by matters of scientific history. And why is scientific history boring for most scientists? It is boring because "a scientist's present thoughts and actions are of necessity shaped by what others have done and thought before him; they are a wavefront of a continuous secular process in which The Past does not have a dignified independent existence of its own. Scientific understanding is the integral of a curve of learning; science therefore in some sense comprehends its history within itself." However, as I shall argue in the next chapter, it is possible to propose quite a different explanation for Watson's failure to give what might have seemed due credit to the discoverers of bacterial transformation—namely, that Avery's discovery in 1944 of the genetic role of DNA was simply "premature."

Medawar next considers the element of luck in Watson's quick rise to world fame at the age of 25. He does not think that "Watson was lucky except in the trite sense in which we are all lucky or unlucky—that there were several branching points in his career at which he might easily have gone off in a direction other than the one he took." Thus, according to Medawar, Watson *was* lucky to have chosen to enter science rather than literary studies, thereby allowing his "precocity and style of genius" to be clever about something important. Watson was also a highly privileged young man, in that he fell in, before he had yet done anything to deserve it, with an "inner circle of scientists among whom information is passed by a sort of beating of tom-toms, while others await the publication of a formal paper in a learned journal. But because it was unpremeditated we can count it to luck that Watson fell in with Francis Crick, who (whatever Watson may have intended) comes out in this book as the dominant figure, a man of very great intellectual powers."

Considered as literature, Medawar classifies *The Double Helix* as the only entry known to him under the rubric Memoirs, Scientific. "As with all good memoirs, a fair amount of it consists of trivialities and idle chatter. Like all good memoirs, it has not been emasculated by considerations of good taste. Many of the things Watson says about the people in his story will offend them, but his own artless candor excuses him, for he betrays in himself faults graver than those he professes to discern in others. *The Double Helix* is consistent in literary structure. . . . There is no philosophizing or psychologizing to obscure our understanding: Watson displays but does not observe himself. Autobiographies, unlike all other works of literature, are part of their own subject matter. Their lies, if any, are lies *of* their authors but not *about* their authors—who (when discovered in falsehood) merely reveal a truth about themselves, namely, that they are liars."

Medawar believes that Watson's book will become a classic, not only in that it will go on being read, but also in that it presents an object lesson of the nature of the creative process in science. As Watson's story shows, that process involves a rapid alternation of "hypothesis and inference, feedback and modified hypothesis. . . . No layman who reads this book with any kind of understanding will ever again think of the scientist as a man who cranks a machine of discovery. No beginner in science will henceforward believe that discovery is bound to come his way if only he practices a certain Method, goes through a well-defined performance of hand and mind."

Before closing this review of reviews, I must report that when Watson sent me his manuscript of *The Double Helix* in the fall of 1966 (then still entitled "Honest Jim"), I urged him not to publish it in its original form. I pointed out to him that I considered it to be of rather low literary quality, and that I thought its gossipy style would preclude its being of interest to anyone not personally acquainted with the protagonists of his story. And so, as has almost always been the case for any difference in opinion we have had in the twenty years we have known each other, Watson proved to be right and I wrong.

*   *   *

Postscript (1978). Medawar's prediction that Watson's book would become a classic was to be borne out. Nearly one million copies were sold in the intervening ten years, and foreign editions appeared in at least seventeen languages, including Latvian, Romanian, and Thai. But, however meritorious *The Double Helix* might be for understanding the sociology of science and the creative process in science, its immense popular success seems unlikely to be accounted for wholly on those grounds. More likely, Bronowski's insight—that Watson's story has to be seen as a classic fairy tale in modern dress—supplies the missing clue. As is well-known to analytical psychology, the classic fairy tale owes its eternal popularity to satisfying deep, subliminal affective needs of its audience. This aspect of *The Double Helix* was, in fact, examined eventually in greater detail by the literary critic William Cadbury in the *Modern Language Quarterly* [31, 474–491 (1970)]. Cadbury found that the book is a second great achievement of Watson's, not merely a record of his first. It is an original and essentially literary statement about an important, and universally interesting, facet of human existence, namely the nature of success. Watson shows what it means and what it takes to be successful as a creative person. In contrast to the inner-directed, "Golden Age" bent for self-fulfillment, of believing that being someone counts for more than knowing something, Watson has little interest in finding out who he is. He knows that there is something out there to be achieved, and that, if only he tries harder, he can do it. Evidently, despite the latter-day preponderance of biographies harping on ethnicity, affirmative action, social responsibility, and relevance, there still remains an audience hungry for Watson's old-fashioned success story of the lone knight measuring himself against immense odds. In the light of Cadbury's analysis, *The Double Helix* can be seen as an inspirational message that gives hope that—the approaching End of Progress notwithstanding—there may still be scope for the exertions of Faustian Man.

Mutual misconceptions of scientists and artists regarding their respective working styles. Top. Scientists' view of the artist at work. Frédéric Chopin (played by Cornel Wilde) sits at his Pleyel pianoforte, and, inspired by his muse George Sand (Merle Oberon), composes his "Preludes." [From the 1945 Columbia Pictures production A Song to Remember.] *Bottom. Artists' view of the scientist at work. Louis Pasteur (Paul Muni) has the sudden inspiration to discover the vaccine for rabies. [From the 1935 Warner Brothers film* The Story of Louis Pasteur. *Both photographs are from the Museum of Modern Art Film Stills Archive.]*

5

PREMATURITY AND UNIQUENESS IN SCIENTIFIC DISCOVERY

[1971]

The fantastically rapid progress of molecular genetics from about 1940 to 1965 eventually obliged merely middle-aged participants in its early development to look back on their early work from a depth of historical perspective that for scientific specialties flowering in earlier times came only after all the witnesses of the first blossoming were long dead. It was as if the late-eighteenth-century colleagues of Joseph Priestley and Antoine Lavoisier had still been active in chemical research and teaching in the 1930s, after atomic structure and the nature of the chemical bond had been revealed. This somewhat depressing personal vantage provided a singular opportunity to assay the evolution of a scientific field. In reflecting on the history of molecular genetics from the viewpoint of my own experience I found that two of its most famous incidents—Oswald Avery's identification of DNA as the active principle in bacterial transformation and hence as genetic material, and Watson and Crick's discovery of the DNA double helix—illuminate two general problems of cultural history. The case of Avery throws light on the question of whether it is meaningful or merely tautologous to say that a discovery is "ahead of its time," or premature. And the case of Watson and Crick can be used, and in fact has been used, to discuss the question of whether there is anything unique in a scientific discovery, in view of the likelihood that if Dr. *A* had not discovered Fact *X* today, Dr. *B* would have discovered it tomorrow.

In line with the burgeoning literary activity of the later 1960s reflecting the onset of the Academic Period, I published a brief retrospective essay on molecular genetics, with particular emphasis on its origins (G. S. Stent, "That Was the Molecular Biology That Was," *Science* 160, 390–395, 1968). In that historical account I mentioned neither Avery's name nor DNA-mediated bacterial transformation. Thus, just as Watson did not in his *The Double Helix*, so did I not trouble to give due credit to the authors of what now is justly seen as a major breakthrough in the understanding of the gene. My essay elicited a letter to the editor by a microbiologist, who complained: "It is a sad and surprising omission that . . . Stent makes no mention of the definitive proof of DNA as the basic hereditary substance by O. T. Avery, C. M. MacLeod and Maclyn McCarty. The growth of [molecular genetics] rests upon this experimental proof. . . . I am old enough to remember the excitement and enthusiasm induced by the publication of the paper by Avery, MacLeod and McCarty. Avery, an effective bacteriologist, was a quiet, self-effacing, non-disputatious gentleman. These characteristics of personality should not [cause] the general scientific public . . . to let his name go unrecognized."

I was taken aback by this letter and replied that I should indeed have mentioned Avery's 1944 first proof that DNA is the hereditary substance. I went on to say, however, that in my opinion it is not true that the growth of molecular genetics rests on Avery's proof. For many years that proof actually had little impact on geneticists. The reason for the delay was not that Avery's work was unknown to or mistrusted by geneticists but that it was "premature."

My *prima facie* reason for saying Avery's discovery was premature is that it was not appreciated in its day. By lack of appreciation I do not mean that Avery's discovery went unnoticed, or even that it was not considered important. What I do mean is that geneticists did not seem to be able to do much with it or build on it. That is, in its day Avery's discovery had virtually no effect on the general discourse of genetics.

This statement can be readily supported by an examination of the scientific literature. For example, a convincing demonstration of the lack of appreciation of Avery's discovery is provided by the 1950 golden jubilee of genetics symposium "Genetics in the 20th Century." In the proceedings of that symposium some of the most eminent geneticists published essays that surveyed the progress of the first 50 years of genetics and assessed its status at that time. Only one of the 26 essayists saw fit to make more than a passing reference to Avery's discovery, then six years old. He was a colleague of Avery's at the Rockefeller Institute, and he expressed some doubt that the active transforming principle was really pure DNA. The then leading philosopher of the gene, H. J. Muller, contributed an essay on the nature of the gene that mentions neither Avery nor DNA.

So why was Avery's discovery not appreciated in its day? Because it was "premature." But is this really an explanation or is it merely an empty tautology? In other words, is there a way of providing a criterion of the prematurity of a discovery other than its failure to make an impact? Yes, there is such a criterion: A discovery is premature if its implications cannot be connected by a series of simple logical steps to canonical, or generally accepted, knowledge.

Why could Avery's discovery not be connected with canonical knowledge? Ever since DNA had been discovered in the cell nucleus by Miescher in 1869 it had been suspected of exerting some function in hereditary processes. This suspicion became stronger in the 1920s, when it was found that DNA is a major component of the chromosomes. The then current view of the molecular nature of DNA, however, made it well-nigh inconceivable that DNA could be the carrier of hereditary information. First, until well into the 1930s DNA was generally thought to be merely a tetranucleotide composed of one unit each of adenylic, guanylic, thymidylic and cytidylic acids. Second, even when it was finally realized by the early 1940s that the molecular weight of DNA is actually much higher than the tetranucleotide hypothesis required, it was still widely believed that the tetranucleotide was the basic repeating unit of the large DNA polymer in which the four units mentioned recur in regular sequence. DNA was therefore viewed as a uniform macromolecule that, like other monotonous polymers such as starch or cellulose, is always the same, no matter what its biological source. The ubiquitous presence of DNA in the chromosomes was therefore generally explained in purely physiological or structural terms. It was usually to the chromosomal protein that the informational role of the genes had been assigned, since the great differences in the specificity of structure that exist between various proteins in the same organism, or between similar proteins in different organisms, had been appreciated since the beginning of the century. The conceptual difficulty of assigning the genetic role to DNA had not escaped Avery. In the conclusion of his paper he stated: "If the results of the present study of the transforming principle are confirmed, then nucleic acids must be regarded as possessing biological specificity the chemical basis of which is as yet undetermined."

By 1950, however, the tetranucleotide hypothesis had been overthrown, thanks largely to the work of Erwin Chargaff. He showed that, contrary to the demands of that hypothesis, the four nucleotides are not necessarily present in DNA in equal proportions. He found, furthermore, that the exact nucleotide composition of DNA differs according to its biological source, suggesting that DNA might not be a monotonous polymer after all. And so when two years later, in 1952, Alfred Hershey and Martha Chase of the Carnegie Institution's laboratory in Cold Spring Harbor, N.Y., showed that on infection of the host bacterium by a bacterial virus at least 80 percent of the viral DNA enters the cell and at least 80 percent of the viral protein remains outside, it was possible to connect their conclusion that DNA is the genetic material with canonical knowledge. Avery's "as yet undetermined chemical basis of the biological specificity of nucleic acids" could now be seen as the precise sequence of the four nucleotides along the polynucleotide chain. The general impact of the Hershey-Chase experiment was immediate and dramatic. DNA was suddenly in and protein was out, as far as thinking about the nature of the gene was concerned. Within a few months there arose the first speculations about the genetic code, and Watson and Crick were inspired to set out to discover the structure of DNA.

Of course, Avery's discovery is only one of many premature discoveries in the history of science. I have presented it here for consideration mainly because of my own failure to appreciate it when I joined Max Delbrück's bacterial virus group at the California Institute of Technology in 1948. Since then I have often wondered what my later career would have been like if I had only been astute enough to appreciate Avery's discovery and infer from it four years before Hershey and Chase that DNA must also be the genetic material of our own experimental organism.

Probably the most famous case of prematurity in the history of biology is associated with the name of Gregor Mendel, whose discovery of the gene in 1865 had to wait 35 years before it was "rediscovered" at the turn of the century. Mendel's discovery made no immediate impact, it can be argued, because the concept of discrete hereditary units could not be connected with canonical knowledge of anatomy and physiology in the middle of the nineteenth century. Furthermore, the statistical methodology by means of which Mendel interpreted the results of his pea-breeding experiments was entirely foreign to the way of thinking of contemporary biologists. By the end of the nineteenth century, however, chromosomes and the chromosome-dividing processes of mitosis and meiosis had been discovered and Mendel's results could now be accounted for in terms of structures visible in the microscope. Moreover, by then the application of statistics to biology had become commonplace. Nonetheless, in some respects Avery's discovery is a more dramatic example of prematurity than Mendel's. Whereas Mendel's discovery seems hardly to have been mentioned by anyone until its rediscovery, Avery's discovery was widely discussed and yet it could not be appreciated for eight years.

Cases of delayed appreciation of a discovery exist also in the physical sciences. One example (as well as an explanation of its circumstances in terms of the concept to which I refer here as prematurity) has been provided by Michael Polanyi on the basis of his own experience. In the years 1914–1916 Polanyi published a theory of the adsorption of gases on solids which assumed that the force attracting a gas molecule to a solid surface depends only on the position of the molecule, and not on the presence of other molecules, in the force field. In spite of the fact that Polanyi was able to provide strong experimental evidence in favor of his theory, it was generally rejected. Not only was the theory rejected, it was also considered so ridiculous by the leading authorities of the time that Polanyi believes continued defense of his theory would have ended his professional career if he had not managed to publish work on more palatable ideas. The reason for the general rejection of Polanyi's adsorption theory was that at the very time he put it forward the role of electrical forces in the architecture of matter had just been discovered. Hence there seemed to be no doubt that the adsorption of gases must also involve an electrical attraction between the gas molecules and the solid surface. That point of view, however, was irreconcilable with Polanyi's basic assumption of the mutual independence of individual gas molecules in the adsorption process. It was only in the 1930s, after a new theory of cohesive molecular forces based on quantum-mechanical resonance rather than on electrostatic attraction had been developed, that it became conceivable that gas molecules could behave in the way Polanyi's experiments indicated they were actually behaving. Meanwhile Polanyi's theory had been consigned so authoritatively to the ashcan of crackpot ideas that it was rediscovered only in the 1950s.

* * *

Still, can the notion of prematurity be said to be a useful historical concept? First of all, is prematurity the only possible explanation for the lack of contemporary appreciation of a discovery? Evidently not. For example, my microbiologist critic suggested that it was the "quiet, self-effacing, non-disputatious" personality of Avery that was the cause of the failure of his contribution to be recognized. Furthermore, in an essay on the history of DNA research Chargaff supports the idea that personal modesty and aversion to self-advertisement account for the lack of contemporary scientific appreciation. He attributes the 75-year lag between Miescher's discovery of DNA and the general appreciation of its importance to Miescher's being "one of the quiet in the land," who lived when "the giant publicity machines, which today accompany even the smallest move on the chess-board of nature with enormous fanfares, were not yet in place." Indeed, the 35-year hiatus in the appreciation of Mendel's discovery is often attributed to Mendel's having been a modest monk living in an out-of-the-way Moravian monastery. Hence the notion of prematurity provides an alternative to the invocation—in my opinion an inappropriate one for the instances mentioned here—of the lack of publicity as an explanation for delayed appreciation.

More important, does the prematurity concept pertain only to retrospective judgments made with the wisdom of hindsight? No, I think it can be used also to judge the present. Some recent discoveries are still premature at this very time. One example of here-and-now prematurity is the alleged finding that experiential information received by an animal can be stored in nucleic acids or other macromolecules.

Some 10 years ago there began to appear reports by experimental psychologists purporting to have shown that the engram, or memory trace, of a task learned by a trained animal can be transferred to a naïve animal by injecting or feeding the recipient with an extract made from the tissues of the donor. At that time the central lesson of molecular genetics—that nucleic acids and proteins are informational macromolecules—had just gained wide currency, and the facile equation of nervous information with genetic information soon led to the proposal that macromolecules—DNA, RNA or protein—store memory. As it happens, the experiments on which the macromolecular theory of memory is based have been difficult to repeat, and the results claimed for them may indeed not be true at all. It is nonetheless significant that few neurophysiologists have even bothered to check these experiments, even though it is common knowledge that the possibility of chemical memory transfer would constitute a fact of capital importance. The lack of interest of neurophysiologists in the macromolecular theory of memory can be accounted for by recognizing that the theory, whether true or false, is clearly premature. There is no chain of reasonable inferences by means of which our present, albeit highly imperfect, view of the functional organization of the brain can be reconciled with the possibility of its acquiring, storing and retrieving nervous information by encoding such information in molecules of nucleic acid or protein. Accordingly for the community of neurophysiologists there is no point in devoting time to checking on experiments whose results, even if they were true as alleged, could not be connected with canonical knowledge.

The concept of here-and-now prematurity can be applied also to the troublesome subject of ESP, or extrasensory perception. In the summer of 1948 I happened to hear a heated argument at Cold Spring Harbor between two future mandarins of molecular biology, Salvador Luria of Indiana University and R. E. Roberts of the Carnegie Institution's laboratory in Washington. Roberts was then interested in ESP, and he felt it had not been given fair consideration by the scientific community. As I recall, he thought that one might be able to set up experiments with molecular beams that could provide more definitive data on the possibility of mind-induced departures from random distributions than J. B. Rhine's then much discussed card-guessing procedures. Luria declared that not only was he not interested in Roberts' proposed experiments but also in his opinion it was unworthy of anyone claiming to be a scientist even to discuss such rubbish. How could an intelligent fellow such as Roberts entertain the possibility of phenomena totally irreconcilable with the most elementary physical laws? Moreover, a phenomenon that is manifest only to specially endowed subjects, as claimed by "parapsychologists" to be the case for ESP, is outside the proper realm of science, which must deal with phenomena accessible to every observer. Roberts replied that far from his being unscientific, it was Luria whose bigoted attitude toward the unknown was unworthy of a true scientist. The fact that not everyone has ESP only means that it is an elusive phenomenon, similar to musical genius. And just because a phenomenon cannot be reconciled with what we now know, we need not shut our eyes to it. On the contrary, it is the duty of the scientist to try to devise experiments designed to probe its truth or falsity.

It seemed to me then that both Luria and Roberts were right, and in the intervening years I often thought about this puzzling disagreement, unable to resolve it in my own mind. Finally six years ago I read a review of a book on ESP by my Berkeley colleague C. West Churchman, and I began to see my way toward a resolution. Churchman stated that there are three different possible scientific approaches to ESP. The first of these is that the truth or falsity of ESP, like the truth or falsity of the existence of God or of the immortality of the soul, is totally independent of either the methods or the findings of empirical science. Thus the problem of ESP is defined out of existence. I imagine that this was more or less Luria's position.

Churchman's second approach is to reformulate the ESP phenomenon in terms of currently acceptable scientific notions, such as unconscious perception or conscious fraud. Hence, rather than defining ESP out of existence, it is trivialized. The second approach probably would have been acceptable to Luria too, but not to Roberts.

The third approach is to take the proposition of ESP literally and to attempt to examine in all seriousness the evidence for its validity. That was more or less Roberts' position. As Churchman points out, however, this approach is not likely to lead to satisfactory results. Parapsychologists can maintain with some justice that the existence of ESP has already been proved to the hilt, since no other set of hypotheses in psychology has received the degree of critical scrutiny that has been given to ESP experiments. Moreover, many other phenomena have been accepted on much less statistical evidence than what is offered for ESP. The reason Churchman advances for the futility of a strictly evidential approach to ESP is that in the absence of a hypothesis of how ESP could work it is not possible to decide whether any set of relevant observations can be accounted for only by ESP to the exclusion of alternative explanations.

After reading Churchman's review I realized that Roberts would have been ill-advised to proceed with his ESP experiments, not because, as Luria had claimed, they would not be "science" but because any positive evidence he might have found in favor of ESP would have been, and would still be, premature. That is, until it is possible to connect ESP with canonical knowledge of, say, electromagnetic radiation and neurophysiology no demonstration of its occurrence could be appreciated.

Is the lack of appreciation of premature discoveries merely attributable to the intellectual shortcoming or innate conservatism of scientists who, if they were only more perceptive or more open-minded, would give immediate recognition to any well-documented scientific proposition? Polanyi is not of that opinion. Reflecting on the cruel fate of his theory half a century after first advancing it, he declared: "This miscarriage of the scientific method could not have been avoided. . . . There must be at all times a predominantly accepted scientific view of the nature of things, in the light of which research is jointly conducted by members of the community of scientists. A strong presumption that any evidence which contradicts this view is invalid must prevail. Such evidence has to be disregarded, even if it cannot be accounted for, in the hope that it will eventually turn out to be false or irrelevant."

That is a view of the operation of science rather different from the one commonly held, under which acceptance of authority is seen as something to be avoided at all costs. The good scientist is seen as an unprejudiced man with an open mind who is ready to embrace any new idea supported by the facts. The history of science shows, however, that its practitioners do not apppear to act according to that popular view.

*　*　*

As was set forth in the preceding chapter, Chargaff wrote one of the many reviews of *The Double Helix*, Watson's autobiographical account of his and Crick's discovery of the structure of DNA. In his review Chargaff found that *"Timon of Athens* could not have been written, 'Les Desmoiselles d'Avignon' not have been painted, had Shakespeare and Picasso not existed. But of how many scientific achievements can this be claimed? One could almost say that, with very few exceptions, it is not the men that make science, it is science that makes the men. What *A* does today, *B* or *C* or *D* could surely do tomorrow."

On reading this passage, I was surprised to find an eminent scientist embracing historicism (the theory championed by Hegel and Marx holding that history is determined by immutable forces rather than by human agency) as an explanation for the evolution of science while at the same time professing belief in the libertarian "great man" view of history for the evolution of art. As can be gathered from my comments about Chargaff's review in the preceding chapter, I suspected at first that Chargaff had made his empty point only to downgrade the importance of Watson and Crick's discovery, since I found it hard to believe that anyone could hold such contradictory, and to me obviously false, views regarding these two most important domains of human creativity. But then I began to ask scientific friends and colleagues whether they too, by any chance, thought there was an important qualitative difference between the achievements of art and of science, namely that the former are unique and the latter inevitable. To my even greater surprise, I found that most of them seemed to agree with Chargaff. Yes, they said, it is quite true that we would not have had *Timon of Athens* or "Les Desmoiselles d'Avignon" if Shakespeare and Picasso had not existed, but if Watson and Crick had not existed, we would have had the DNA double helix anyway. Therefore, contrary to my first impression, it does not seem to be all that obvious that this proposition has little philosophical or historical merit. Hence I shall now attempt to show that there is no such profound difference between the arts and sciences in regard to the uniqueness of their creations.

Before discussing the proposition of differential uniqueness of creation it is necessary to make an explicit statement of the meaning of "art" and of "science." As I pointed out in Chapter 2, the traditionalist view of the arts and the sciences is that both are activities that endeavor to discover and communicate truths about the world. The domain to which the artist addresses himself is the inner, subjective world of the emotions. Artistic statements therefore pertain mainly to relations between private events of affective significance. The domain of the scientist, in contrast, is the outer, objective world of physical phenomena. Scientific statements therefore pertain mainly to relations between or among public events. Thus the transmission of information and the perception of meaning in that information constitute the central content of both the arts and the sciences. A creative act on the part of either an artist or a scientist would mean his formulation of a new meaningful statement about the world, an addition to the accumulated capital of what is sometimes called "our cultural heritage." Let us therefore examine the proposition that only Shakespeare could have formulated the semantic structures represented by *Timon*, whereas people other than Watson and Crick might have made the communication represented by their paper, "A Structure for Deoxyribonucleic Acid," published in *Nature* in the spring of 1953.

First, it is evident that the exact word sequence that Watson and Crick published in *Nature* would not have been written if the authors had not existed, any more than the exact word sequence of *Timon* would have been written without Shakespeare, at least not until the fabulous monkey typists complete their random work at the British Museum. And so both creations are from that point of view unique. We are not really concerned, however, with the exact word sequence. We are concerned with the content. Thus we admit that people other than Watson and Crick would eventually have described a satisfactory molecular structure for DNA. But then the character of *Timon* and the story of his trials and tribulations not only might have been written without Shakespeare but also were written without him. Shakespeare merely reworked the story of *Timon* he had read in William Painter's collection of classic tales, *The Palace of Pleasure*, published 40 years earlier, and Painter in turn had used as his sources Plutarch and Lucian. But then we do not really care about Timon's story; what counts are the deep insights into human emotions that Shakespeare provides in his play. He shows us here how a man may make his response to the injuries of life, how he may turn from lighthearted benevolence to passionate hatred toward his fellow men. Can one be sure, however, that *Timon* is unique from this bare-bones standpoint of the work's artistic essence? No, because who is to say that if Shakespeare had not existed no other dramatist would have provided for us the same insights? Another dramatist would surely have used an entirely different story (as Shakespeare himself did in his much more successful *King Lear*) to treat the same theme and he might have succeeded in pulling it off. The reason no one seems to have done it since is that Shakespeare had already done it in 1607, just as no one discovered the structure of DNA after Watson and Crick had already discovered it in 1953.

Hence we are finally reduced to asserting that *Timon* is uniquely Shakespeare's, because no other dramatist, although he might have brought us more or less the same insights, would have done it in quite the same exquisite way as Shakespeare. But here we must not shortchange Watson and Crick and take for granted that those other people who eventually would have found the structure of DNA would have found it in just the same way and produced the same revolutionary effect on contemporary biology. On the basis of my acquaintance with the personalities then engaged in trying to uncover the structure of DNA, I believe that if Watson and Crick had not existed, the insights they provided in one single package would have come out much more gradually over a period of many months or years. Dr. B might have seen that DNA is a double-strand helix, and Dr. C might later have recognized the hydrogen bonding between the strands. Dr. D later yet might have proposed a complementary purine-pyrimidine bonding, with Dr. E in a subsequent paper proposing the specific adenine-thymine and guanine-cytosine nucleotide pairs. Finally, we might have had to wait for Dr. G to propose the replication mechanism of DNA based on the complementary nature of the two strands. All the while Drs. H, I, J, K and L would have been confusing the issue by publishing incorrect structures and proposals. Thus I fully agree with the judgment offered by Peter Medawar in his review of *The Double Helix* that the great thing about Watson and Crick's discovery was "its completeness, its air of finality." As reported in the preceding chapter, Medawar thought that "if Watson and Crick had been seen groping toward an answer, . . . if the solution had come out piecemeal instead of in a blaze of understanding, then it would still have been a great episode in biological history." But it would not have been the dazzling achievement in the "grand romantic manner" that it, in fact, was.

* * *

Why is it that so many scientists apparently fail to see that it can be said of both art and science that whereas "what A does today, B or C or D could surely do tomorrow," B or C or D might nevertheless not do it as well as A, in the same "grand romantic manner." I think a variety of reasons can be put forward to account for this strange myopia. The first of them is simply that most scientists are not familiar with the working methods of artists. They tend to picture the artist's act of creation in the terms of Hollywood: Cornel Wilde in the role of the one and only Frédéric Chopin gazing fondly at Merle Oberon as his muse and mistress George Sand and then sitting down at the Pleyel pianoforte to compose his "Preludes." As scientists know full well, science is done quite differently: Dozens of stereotyped and ambitious researchers are slaving away in as many identical laboratories, all trying to make similar discoveries, all using more or less the same knowledge and techniques, some of them succeeding and some not. Artists, on the other hand, tend to conceive of the scientific act of creation in equally unrealistic terms: Paul Muni in the role of the one and only Louis Pasteur, who while burning the midnight oil in his laboratory has the inspiration to take some bottles from the shelf, mix their contents and thus discover the vaccine for rabies. Artists, in turn, know that art is done quite differently: Dozens of stereotyped and ambitious writers, painters and composers are slaving away in as many identical garrets, all trying to produce similar works, all using more or less the same knowledge and techniques, some succeeding and some not.

A second reason is that the belief in the inevitability of scientific discoveries appears to derive support from the often-told tales of famous cases in the history of science where the same discovery was made independently two or more times by different people. For instance, the independent invention of the calculus by Leibniz and Newton or the independent recognition of the role of natural selection in evolution by Wallace and Darwin. As the study of such "multiple discoveries" by Robert Merton of Columbia University has shown, however, on detailed examination they are rarely, if ever, identical. The reason they are said to be multiple is simply that in spite of their differences one can recognize a semantic overlap between them that is transformable into a congruent set of ideas.

The third, and somewhat more profound, reason is that whereas the cumulative character of scientific creation is at once apparent to every scientist, the similarly cumulative character of artistic creation is not. For instance, it is obvious that no present-day working geneticist has any need to read the original papers of Mendel, because they have been completely superseded by the work of the past century. Mendel's papers contain no useful information that cannot be better obtained from any modern textbook or the current genetical literature. In contrast, the modern writer, composer or painter still needs to read, listen or look at the original works of Shakespeare, Bach or Leonardo, which, so it is thought, have not been superseded at all. In spite of the seeming truth of this proposition, it must be said that art is no less cumulative than science, in that artists no more work in a traditionless vacuum than scientists do. Artists also build on the work of their predecessors; they start with and later improve on the styles and insights that have been handed down to them from their teachers, just as scientists do. To stay with our main example, Shakespeare's *Timon* has its roots in the works of Aeschylus, Sophocles and Euripides. It was those authors of Greek antiquity who discovered tragedy as a vehicle for communicating deep insights into affects, and Shakespeare, drawing on many earlier sources, finally developed that Greek discovery to its ultimate height. To some limited extent, therefore, the plays of the Greek dramatists have been superseded by Shakespeare's. Why, then, have Shakespeare's plays not been superseded by the work of later, lesser dramatists?

Here we finally do encounter an important difference between the creations of art and of science, namely the feasibility of paraphrase. The semantic content of an artistic work—a play, a cantata or a painting—is critically dependent on the exact manner of its realization; that is, the greater an artistic work is, the more likely it is that any omissions or changes from the original detract from its content. In other words, to paraphrase a great work of art—for instance to rewrite *Timon*—without loss of artistic quality requires a genius equal to the genius of the original creator. Such a successful paraphrase would, in fact, constitute a great work of art in its own right. The semantic content of a great scientific paper, on the other hand, although its impact at the time of publication may also be critically dependent on the exact manner in which it is presented, can later be paraphrased without serious loss of semantic content by lesser scientists. Thus the simple statement "DNA is a double-strand, self-complementary helix" now suffices to communicate the essence of Watson and Crick's great discovery, whereas "A man responds to the injuries of life by turning from lighthearted benevolence to passionate hatred toward his fellow men" is merely a platitude and not a paraphrase of *Timon*. It took the writing of *King Lear* to paraphrase (and improve on) *Timon*, and indeed the former has superseded the latter in the Shakespearean dramatic repertoire.

The fourth, and probably deepest, reason for the apparent prevalence among scientists of the proposition that artistic creations are unique and scientific creations are not can be attributed to a contradictory epistemological attitude toward the events in the outer and the inner world. The outer world, which science tries to fathom, is often viewed from the standpoint of materialism, according to which events and the relations between them have an existence independent of the human mind. Hence the outer world and its scientific laws are simply there, and it is the job of the scientist to find them. Thus going after scientific discoveries is like picking wild strawberries in a public park: the berries A does not find today B or C or D will surely find tomorrow. At the same time, many scientists view the inner world, which art tries to fathom, from the standpoint of idealism, according to which events and relations between them have no reality other than their reflection in human thought. Hence there is nothing to be found in the inner world, and artistic creations are cut simply from whole cloth. Here B or C or D could not possibly find tomorrow what A found today, because what A found had never been there. It is not altogether surprising, of course, to find this split epistemological attitude toward the two worlds, since of these two antithetical traditions in Western philosophical thought, materialism is obviously an unsatisfactory approach to art and idealism an unsatisfactory approach to science.

* * *

In recent years more or less contemporaneously with the growth of molecular biology, a resolution of the age-old epistemological conflict of materialism v. idealism was found in the form of what has come to be known as structuralism. Structuralism emerged simultaneously, independently and in different guises in several diverse fields of study, for example in psychology, linguistics, anthropology and biology.

Both materialism and idealism take it for granted that all the information gathered by our senses actually reaches our mind; materialism envisions that thanks to this information reality is mirrored in the mind, whereas idealism envisions that thanks to this information reality is constructed by the mind. Structuralism, on the other hand, has provided the insight that knowledge about the world enters the mind not as raw data but in already highly abstracted form, namely as structures. In the

preconscious process of converting the primary data of our experience step by step into structures, information is necessarily
lost, because the creation of structures, or the recognition of
patterns, is nothing else than the selective destruction of information. Thus since the mind does not gain access to the full set
of data about the world, it can neither mirror nor construct reality. Instead for the mind reality is a set of structural transforms of
primary data taken from the world. This transformation process
is hierarchical, in that "stronger" structures are formed from
"weaker" structures through selective destruction of information. Any set of primary data becomes meaningful only after a
series of such operations has so transformed it that it has become congruent with a stronger structure preexisting in the
mind. As we shall see in Chapter 8, neurophysiological studies
carried out in recent years on the process of visual perception in
higher mammals have not only shown directly that the brain
actually operates according to the tenets of structuralism but also
offer an easily understood illustration of those tenets.

Finally, we may consider the relevance of structuralist philosophy for the two problems in the history of science under discussion here. As far as prematurity of discovery is concerned,
structuralism provides us with an understanding of why a discovery cannot be appreciated until it can be connected logically
to contemporary canonical knowledge. In the parlance of structuralism, canonical knowledge is simply the set of preexisting
"strong" structures with which primary scientific data are made
congruent in the mental-abstraction process. Hence data that
cannot be transformed into a structure congruent with canonical
knowledge are a dead end; in the last analysis they remain
meaningless. That is, they remain meaningless until a way has
been shown to transform them into a structure that is congruent
with the canon.

As far as uniqueness of discovery is concerned, structuralism
leads to the recognition that every creative act in the arts and
sciences is both commonplace and unique. On the one hand, it
is commonplace in the sense that there is an innate correspondence in the transformational operations that different individuals perform on the same primary data. With reference to science,
cognitive psychology has taught that different individuals recognize the same "chairness" of a chair because they all make a

given set of sense impressions from the outer world congruent with the same *Gestalt*, or mental structure. With reference to art, analytic psychology has taught that there is a sameness in the subconscious life of different individuals because an innate human archetype causes them to make the same structural transformations of the events of the inner world. And with reference to both art and science structuralist linguistics has taught that communication between different individuals is possible only because an innate human grammar causes them to transform a given set of semantic symbols into the same syntactic structure. On the other hand, every creative act is unique in the sense that no two individuals are quite the same and hence never perform exactly the same transformational operations on a given set of primary data. Although all creative acts in both art and science are therefore both commonplace and unique, some may nonetheless be more unique than others.

*   *   *

Postscript (1978). To my surprise, the musicologist Leonard B. Meyer, on whose ideas I had relied so extensively in my treatment of the arts in Chapter 2, published a highly critical discussion of the second part of this essay [*Critical Inquiry*, 1, 163–217 (1974)]. Meyer, like Chargaff [and, it should be noted, also Crick *(Nature*, 248, 776–769 (1974)] turned out to find self-evident merit in the proposition that whereas we would not have had *Timon of Athens* had Shakespeare not existed, we would have had the DNA double helix even if Watson and Crick had not existed. In fact, Meyer saw in my rejection of that proposition an attempt to perform (with the help of C. P. Snow) a "shotgun marriage" between the "so-called Two Cultures." According to him, I was trying to force an unnatural union by claiming that both the arts and the sciences are activities that endeavor to discover and communicate truths about the world. As far as discovery is concerned, Meyer held that only scientists discover truths; but they do not create anything. Why? Because "we assume, evidently on good grounds, that while our theories explaining nature may change, the principles governing relationships in the natural world are constant with respect to both time and place." Artists, by contrast, do not discover anything: they create their works, which have no prior existence.

Moreover, the concept of "truth" is not applicable to art, because there are no imaginable data or experiments that could test the validity of a work of art. Great works of art merely "command our assent. Like validated [scientific] theories, they seem self-evident and incontrovertible, meaningful and necessary, infallible and illuminating. There is, without doubt, an aura of 'truth' about them." Science and art are not marriagable because, according to Meyer, scientific theories are *propositional* and works of art are *presentational*.

In a reply to Meyer [*Critical Inquiry*, 1, 683–694, (1975)] I pointed out that in so far as any scientific theory is an abstraction, rather than a mere reflection of the "natural world," it *is* a creation as much as it is a discovery: here the creative act consists of selecting for attention a particular, interesting subset of the near-infinitude of events that the "natural world" continually presents to our consciousness. Second, as regards the concept of "truth," I observed that Meyer confounded the semantic problem of its meaning with the epistemological problem of the validation of scientific propositions. When used in ordinary discourse, "truth" refers to a statement which a person believes to be the case, finds to be in harmony with his internalized picture of the world. So he can believe a statement to be true, even if he lacks all means for its objective validation. Commanding our assent is a sufficient truth condition, which as Meyer admits, *is* met by great works of art. Finally, and most regrettably, I found that Meyer too confounded the essential difference between a work and its semantic content. Obviously a scientific theory (to which the concept of truth is applicable) is not a "work" of science; it is the content of a work, such as a treatise, an article, or a lecture. And a scientific treatise, article, or lecture is just as "presentational" as a work of art. Conversely, the concept of truth is not applicable to a work of art (e.g., to a presentation of *Timon*) but only to its semantic content. Nevertheless, when we compare *how* the contents of the works of art and of science are presented, we *do* find an essential difference: the content of a scientific work is stated explicitly in linguistic terms, whereas the content of a work of art is merely implicit in its structure. This presentational difference may have enormous implications, but the proposition that only works of art, but not works of science, are unique is not one of them.

Jacques Monod. Portrait by Efraim Racker, 1947.
[By permission of Efraim Racker.]

6

MOLECULAR BIOLOGY AND METAPHYSICS

[*1971* and *1973*]

And now the announcement of Watson and Crick about DNA. This is for me the real proof of the existence of God.

Salvador Dali, 1964

In *The Double Helix* Watson merely told his own story, leaving for the reader to draw for himself any cosmic implications about scientists and their work. But two other principal architects of the edifice of molecular genetics, Francis Crick and Jacques Monod, moved into the salon with essays explicitly expounding the profound philosophical significance of the achievements of molecular genetics, not only for the understanding of life processes but also for the fathoming of man's very relation to the Universe.

Crick's essay, entitled *Of Molecules and Men* (University of Washington Press, Seattle, 1966), is prefaced with the above quotation from Dali, but neither the surrealist painter nor God is mentioned in the body of his book. Why then does Crick quote Dali? Although *The Double Helix* opens with Honest Jim's finding: "I have never seen Francis Crick in a modest mood," not even Watson would claim that Crick really believes that in 1953 they delivered final proof for the existence of God. No, Crick evidently finds Dali's statement a tremendous joke, and though Dali's intent was surely serious, Crick is making fun of him, by according Dali the place of honor under the masthead of an antireligious tract.

Far from having proven the existence of God, so Crick intimates, the achievements of molecular genetics have made religious beliefs even more superfluous and outdated than they had been before the structure of DNA was found. For instance, vitalism, the eighteenth-century doctrine which held that the phenomena of life can only be explained by a mysterious "vital force" and which Crick identifies with Christianity, and especially with Catholicism, has now, at last, been definitely smashed. *Of Molecules and Men* lies squarely in the philosophical tradition of positivism, which stoutheartedly dismisses as nonsensical all concepts, such as "soul," which cannot be the subject of operationally meaningful explications. In line with that tradition, Crick identifies himself with the social and moral philosophy of scientism, by recommending that "science in general, and natural selection in particular, should become the basis on which we are to build the new culture."

Later in this and other chapters I shall try to show that Dali actually sized up the situation quite correctly: the discovery of the DNA structure *has* furnished proof for the existence of God. Denials by adherents of positivism notwithstanding, the metaphysical axiom of Western science, namely that the phenomena of the world are accessible to analysis by human reason, has its rational roots in the belief in the existence of God-given Natural Law. And thus by their brilliant contribution to the explanation of life processes in terms of ordinary physics and chemistry, Watson and Crick have made it more difficult, rather than less difficult, to abandon belief in God.

Monod's *Chance and Necessity* (Knopf, New York, 1971) covers some of the same ground (e.g., the molecular genetic interment of vitalism) and dismisses religion as summarily as does *Of Molecules and Men*. But Monod's philosophical scope is broader and his arguments are more fully developed than Crick's. Monod worked in Paris at the Pasteur Institute, whose Director he was at the time of his death in 1976. Monod and his colleague

François Jacob played a crucial role in the formulation of the network of molecular genetic theories of the Dogmatic Period, in that the eventual understanding of the regulation of gene function came mainly through the ideas of these two French biologists. Monod was very much a man of the world: a politically engaged person, leader in the wartime French Resistance, accomplished musician, elegant writer and, in his last years, probably the best-known scientist in France. Hence a statement of the philosophical insights reached by this extraordinary man could have been expected to be of interest to a public much wider than his scientific colleagues. And indeed, Monod's book became much more widely known than Crick's. In France, *Chance and Necessity* achieved instant success, where it remained on the best-seller list for many weeks, ranking just behind the French translation of *Love Story*. It was also highly successful in Germany and other Continental countries, but rather less so in Britain and the United States, where the literate public has much less affective involvement with Monod's twin targets of attack: Judeo-Christianity and Marxism.

* * *

Monod begins what he subtitles his "Essay on the Natural Philosophy of Modern Biology" by drawing attention to three general properties that characterize living beings and distinguish them from the rest of the universe. The first of these is *teleonomy*, which means that living beings are objects endowed with a purpose. The second general property is *autonomous morphogenesis*, which means that living beings are self-constructing machines. And the third general property is *reproductive invariance*, which means that among living beings, like begets like. Of these three general properties, scientific analysis can more readily fathom the second and third than the first. For however amazing may seem on first sight the existence of objects capable of self-construction and faithful self-reproduction, there is no longer any reason to doubt that a complete account can be given of these faculties. Indeed, molecular biology has now gone very far toward providing a full physico-chemical explanation of biological self-construction and self-reproduction, and Monod devotes about one third of his essay to giving a summary of the studies carried out during the past quarter century on proteins and nucleic acids—on enzymes and genes and their relation via the genetic code—that have shown how cells transact the business of life. But when we try to deal with biological purpose we run into trouble because, as Monod points out, attribution of purpose to any natural object involves us in a contradiction with what he calls the *principle of objectivity*. For "the cornerstone of scientific method is the postulate that nature is objective. In other words, the *systematic* denial that 'true' knowledge can be got at by interpreting phenomena in terms of final causes—that is to say, of 'purpose.' " Thus while the purposive character of life is *prima facie* apparent, scientific objectivity obliges us to deny it. "This self-same contradiction is in fact the central problem of biology."

How have philosophers of the past tried to deal with this problem? Mainly by positing some general or cosmic principle that drives creation and of which any purposive character or behavior is but a particular manifestation. And Monod assigns such philosophers to two general schools, the *vitalist* and the *animist*. The vitalist school, of which "there has probably been no more illustrious proponent than Henri Bergson," believes that the general teleonomic principle, the *élan vital*, pertains only to life. Indeed it is this mysterious life-force that sets apart living from nonliving matter.

The animist school, by contrast, projects into *all* nature, living and nonliving, "man's awareness of the intensely teleonomic functioning of his own central nervous system. Animism is, in other words, the hypothesis that natural phenomena can and must be explained in the same manner, by the same 'laws,' as subjective human activity, conscious and purposive." The animist hypothesis must be of very high antiquity and probably came into being when early man began to formulate his very first philosophy of nature. In those faraway days "animism established a covenant between man and nature, a profound alliance outside of which seems to stretch only terrifying solitude."

Though nowadays animism is usually thought of only in connection with the beliefs of primitive peoples of the Australian bush or the headwaters of the Amazon, two very popular modern European philosophies of nature are nothing but animism.

One of these is the Christian evolutionary theory of Teilhard de Chardin that posits an "ascending energy" in the cosmos leading to ever higher things. Monod is "struck by the intellectual spinelessness of this philosophy," which "would not merit attention but for the startling success it has encountered even in scientific circles." And the other animist holdover is the dialectical materialism of Marx and Engels, particularly its "vulgate" version, which has been guiding Communist philosophical thought for the past fifty years. Dialectical materialism asserts that the universe is in a state of perpetual evolution because movement is inherent in matter. This movement embodies a dialectic of contradictions, and since out of these contradictions there arise new and better things, evolution necessarily leads to progress. Monod finds that dialectical materialism is in a state of "epistemological bankruptcy," not only because it violates the principle of objectivity but also because it has been an obvious hindrance, rather than an aid, to the progress of science.

* * *

How *is* the dilemma of biological purpose to be resolved? Monod believes "that we can assert today that a universal theory, however completely successful in other domains, could never encompass the [living world], its structure, and its evolution as phenomena *deductible* from first principles." Instead of searching for a universal theory we must seek the explanation of life's purpose in *chance*. That is to say, the living world does not contain a predictable class of objects or of events but constitutes a particular occurrence, compatible indeed with first principles, but not deductible from those principles, and therefore essentially unpredictable. That is a bitter pill to swallow, since "we would like to think ourselves necessary, inevitable, ordained from all eternity. All religions, nearly all philosophies, and even a part of science testify to the unwearying, heroic effort of mankind in desperately denying its own contingency."

But how can purpose arise from chance? Charles Darwin provided the answer: through an evolutionary process in which not "life-force," not "ascending energy," not "matter in motion," but *natural selection* operating on random variations creates purposeful structures. So what is new? Surely everybody, including

Bergson, Teilhard, and Marx and Engels, has known all about Darwin's theory since it was proposed more than a century ago. What *is* new, according to Monod, is that the recent discoveries of molecular biology have finally resolved a logical contradiction hidden in the Darwinian theory, namely how, despite their breeding true, organisms can constantly sport variants which, if natural selection favors them, once more breed true to give rise to a new line of variant offspring. We now know that organisms carry their hereditary information encased in DNA molecules and that reproductive invariance is but a consequence of the capacity of DNA for faithful self-replication, and of the capacity for self-assembly of the other cell constituents over whose synthesis the DNA molecules preside. Hereditary variants, or *mutants*, arise merely as occasional random imperfections of the DNA replication process. "And so one may say that [this] source of fortuitous perturbations, of 'noise,' which in a nonliving (i.e. nonreplicative) system would lead little by little to disintegration of all structure, is the progenitor of evolution in the [living] world and accounts for its unrestricted liberty of creation, thanks to the replicative structure of DNA: that registry of chance, that tone-deaf conservatory where the noise is preserved along with the music."

After discussing the present frontiers of knowledge in evolutionary biology, particularly the riddles of the origin of life itself, of the higher nervous system, and of such peculiarly human attributes as language from the purview of molecular Darwinism, Monod finally spells out the philosophical implications for the actual human condition of this solution to the problem of biological purpose. He is *not* concerned with the specter of deliberate molecular biological manipulation of the human hereditary capital, or "genetic engineering," the possibility of which Monod declares to be "an illusion, spread by a few superficial minds." No, his concern is the modern soul, which he perceives to be in distress. Now that man finds himself to be nothing more than the result of a series of chance errors in the replicative history of DNA molecules, the millennia-old animist covenant between man and nature, which "discloses the meaning of man by assigning him a necessary place in nature's scheme" and thus satisfies his innate desire for a complete explanation of the cosmos, has been dissolved. And the replacement for that covenant—namely adherence to *objectivism*, or belief that "objective knowledge is the *only* authentic source of truth" leaves "nothing in place of that precious bond [between man and nature] but an anxious quest in a frozen universe of solitude."

So our first order of business is to prepare a thorough, objective revision of our ethical premises, since the values to which man has been, and is still, beholden are based on the old animist covenant. Thus, "for their moral basis the 'liberal' societies of the West still teach—or pay lip service to—a disgusting farrago of Judeo-Christian religiosity, scientistic progressism, belief in the 'natural' rights of man and utilitarian pragmatism." How shall we go about making that revision? *Not* by adopting "the position once and for all that objective truth and theory of value constitute eternally separate, mutually impenetrable domains." Monod finds *that* position "absolutely mistaken," despite its having been "taken by a great number of modern thinkers, whether writers or philosophers, or indeed scientists," and despite his own hardly different view that "knowledge in itself is exclusive of all value judgment . . . whereas ethics, in essence *nonobjective*, is forever barred from the sphere of knowledge." Although the principle of objectivity "prohibits any confusion of value judgments with judgments arrived at through knowledge, these two categories inevitably unite in the form of action, discourse included." What is now needed, therefore, is *authentic* discourse or action, which combines objective truth and values and yet manages to preserve the distinction between these two categories. This is to be contrasted with "*inauthentic* discourse where the two categories are jumbled [and which] can lead only to the most pernicious nonsense." Having mastered the art of authentic discourse one realizes that adherence to the principle of objectivity "*constitutes an ethical* choice and not a judgment arrived at from knowledge," that is to say, is an expression of what Monod calls *the ethic of knowledge*.

* * *

What is the difference between the new ethic of knowledge and the old ethic of animism? The animist ethic claims to descend from immanent laws which assert themselves over man, whereas it is man in his terrible solitude who prescribes the ethic of knowledge for himself.

And what would be some concrete examples of the changes in values wrought by adherence to the ethic of knowledge and abandonment of Judeo-Christian-scientistic pragmatism or Marxist dialectical materialism? Monod does not provide any examples, possibly for the very good reason that the ethic of knowledge seems to let one go right on believing in what the animists have believed in all the while. Judeo-Christians can find comfort in the thought that "as for the highest human qualities, courage, altruism, generosity, creative ambition, the ethic of knowledge both recognizes their sociobiological origin and affirms their transcendent value in the service of the ideal it defines." And Marxists must find some solace in the conclusion that socialism, that great vision of the nineteenth century which "to the young in spirit . . . continues to beckon with grievous intensity," can finally be built upon the ethic of knowledge, after the betrayals it has suffered and the crimes committed in its name. For in his final paragraph Monod declares socialism to be "the conclusion to which the search for authenticity necessarily leads. The ancient covenant is in pieces: man knows at last that he is alone in the universe's unfeeling immensity, out of which he emerged only by chance. His destiny is nowhere spelled out, nor is his duty. The kingdom above or the darkness below: it is for him to choose."

*　*　*

I think this is an important book, mainly because I agree with what Monod calls in the preface his only excuse for it: "the duty which more forcibly than ever thrusts itself upon scientists to apprehend their discipline within the larger framework of modern culture, with a view to enriching the latter not only with technical findings but also with what they may feel to be humanly significant ideas arising from their area of special concern. The very ingenuousness of a fresh look at things (and science possesses an ever-youthful eye) may sometimes shed new light upon old problems."

But because of the very importance of this philosophical statement by one of the major scientific figures of our time, it must be subjected to critical analysis, as no doubt (in line with the beliefs he offers here) Monod himself would have been the first to wish it to be. Alas, I find it difficult to make any such analysis of the last, and for Monod possibly the most important, part of his essay: the promulgation of his "ethics of knowledge." For my feelings about this attempt to provide an "objective" basis for ethical values are more or less those which Monod expresses for the work of Henri Bergson, namely that it is written in "an engaging style and a metaphorical dialectic bare of logic but not of poetry." As far as I do understand it, the ethic of knowledge seems to be, if not outright scientism, a kind of underground crypto-scientism. In any case, Catholic as well as Marxist philosophers of France have already risen to the challenge of seeing their beliefs and ideas summarily consigned to the ash can and have made vigorous counterpoetical attacks on Monod's ethic of knowledge. And Monod's account of the molecular biological explanation of life's self-organization and self-replication *needs* no critical analysis. No one who is familiar with Monod's scientific writings will be surprised to learn that he has provided here for a general audience an excellent summary of the field in whose development he has loomed so large. Indeed, he has managed to view from interesting new perspectives such seemingly workaday phenomena as enzyme reaction and protein structure.

But when it comes to what I consider to be the philosophical core of this essay, namely objectivity and the end of animism, critical comment both is needed and can be readily given. Like Monod, I believe that an age-old covenant between man and nature is being dissolved in our days and that this dissolution signals an important milestone in the evolution of the intellect. However, I find that Monod has not correctly described either the implications for science of that covenant or what is about to replace it. Unfortunately, Monod fails to provide explicit definitions of many of his key concepts—of "science," "objectivity," "solitude," or "true knowledge"—which is troublesome because internal evidence suggests that he endows some of these terms with unconventional meanings. For instance, the only clue he gives to his meaning of "science" occurs in the preface of his essay, where he declares his belief that "the ultimate aim of the whole of science is indeed to clarify man's relationship to the universe." However, in ordinary philosophical parlance *that* aim happens to be the very goal of *metaphysics*, a branch of knowledge which Monod obviously believes to be in conflict with the principle of objectivity and for which, in contrast to science, he has little use.

So let us begin our analysis with animism, which Monod does define as the hypothesis that natural phenomena can and must be explained by the same laws as conscious and purposive human activity. (Here is another example of Monod's unconventional use of philosophical terms, since "animism" is generally understood to designate a more specific hypothesis, namely the belief that all natural objects have a soul, or *anima*.) Certainly the most basic of the laws projected by man into nature is *causality*, or the belief that the events he observes in the outer world resemble his own conscious acts in their being connected as cause and effect, rather than occurring haphazardly. And since the belief in causality, or in the orderliness of events, is the fundament on which almost all of man's past attempts at analyzing nature were based, science, as I understand that term, far from being incompatible with what Monod means by animism, is one of its most important manifestations. Indeed, even the most elementary dimensions in terms of which scientists attempt to describe the very events that causality is supposed to connect, such as time, space, mass, and temperature, are nothing more than projections into nature of man's own physiology and anatomy.

* * *

Now what does Monod mean by "objectivity"? If he intends to imply freedom from *any* subjective projections into nature, then the postulate that nature is objective cannot be, as Monod insists it is, the "cornerstone of the scientific method," and the principle of objectivity cannot be "consubstantial" with science. However, these two assertions *could* be true if "objectivity" were to designate a more limited stricture, namely the refusal to posit any cause-effect relation between events, such as purpose, that observation cannot in principle either prove or disprove. This was more or less the principle of objectivity which the philosophers of positivism held "consubstantial" with science. Like

Monod, the positivists disdained metaphysics, but unlike him they found problems such as man's relationship to the universe to be pure nonsense.

But if Monod does intend the positivistic meaning of objectivity, then a major part of his essay, namely the explanation of the origin of biological purpose in chance rather than necessity, would fail the test of objectivity. First of all, Monod does not seem to have noticed that Darwin's principle of natural selection, on which he places such a heavy reliance, is not itself an objective scientific proposition. The essence of the theory of natural selection is the concept of *fitness:* the fitter an organism, the higher its rate of reproduction relative to other organisms in its environment. Since the criterion of fitness is reproduction, there is no conceivable observation or experiment that could lead one to the conclusion that evolution has *not* proceeded by a process in which the fittest organisms reproduce most prolifically—no more than, as Monod points out, it is possible "to imagine an experiment which could prove the *non-existence* anywhere in nature of a purpose. This critique is not to suggest that the principle of natural selection is false; on the contrary, it is logically but not "objectively" true.

It so happens that a controversy currently agitating the students of evolution illustrates the nonobjectivity of the natural selection idea. In the past two or three years, probably after Monod completed his original manuscript, some biologists have inferred from comparative analyses of the detailed molecular structure of particular proteins of presently living organisms, from yeast, worms, and insects to fish, frogs, fowl, and man, that natural selection cannot have played the all-important role attributed to it by Darwin. These non-Darwinian molecular evolutionists assert that most of the genetic mutations that were responsible for the present differences in protein structure between living species represented *neutral* changes. In other words, in denial of the point central to Monod's argument that these very molecular changes account for the genesis of purposeful structures, the bulk of historical mutations did not affect the function of proteins at all. This claim of the non-Darwinians is being hotly denied by the Darwinian True Believers, who insist that no mutation which survived the rigors of evolutionary selection *can have been neutral*. In this dispute, the non-Darwinians are obviously fighting a losing battle, since *by definition* the test of the superior fitness of the historical mutations is the fact that they are now carried by modern organisms which have reproduced more prolifically than those less fortunate ancient creatures that have fallen by the evolutionary wayside.

In addition to its reliance on natural selection, Monod's argument in favor of chance rather than necessity as the origin of biological purpose harbors a second nonobjective proposition. That is his claim that the biological innovations on which evolution feeds arise by chance. Here Monod seems to have overlooked the point that chance origin is part and parcel of the very concept of innovation. From a preordained universe, such as "Laplace's world from which chance is excluded by definition," innovation would also be automatically excluded. That is, since in a totally deterministic world the whole future is immanent in its initial conditions, nothing really new *could* ever arise. Hence the philosophical problem at issue is not whether or not innovation arises by chance but whether innovation is *possible*, a metaphysical question (whose affirmative answer Monod, of course, takes for granted) that cannot be settled by recourse to any principle of objectivity.

From the viewpoint of natural selection, adherence to the animist covenant contributed to man's evolutionary fitness, since by making science possible it has allowed man to dominate nature. In our time, the ancient covenant is finally breaking up, not, I think, because as Monod declares, after so many thousands of years it has finally dawned on people that "objective knowledge is the *only* authentic source of truth" but because science has finally shown that in nature there *is* no "truth." Or, out there, as Gertrude Stein once said of Oakland, California, there is no there there.

Scientific subversion of the covenant began to become a serious matter at the turn of this century when physics had evolved to the stage at which problems could be studied involving either tiny subatomic or immense cosmic events on scales of time, space, and mass billions of times smaller or larger than the scales on which man originally projected these basic dimensions into nature. Then it was first found that phenomena on very large or very small scales could not be given an adequate description in ordinary, everyday language and that time, space, and mass had to be denatured from their original, wholly intuitive meanings into formal concepts that run counter to common sense. Soon thereafter it appeared also that even the notion of cause and effect was not a useful one for giving account of events at the subatomic level.

Science, while gaining much greater scope by abandoning ideas such as that mass is conserved, that space is straight, that time is absolute, and that events are connected, was to pay a heavy price for this contravention of common sense. Albert Einstein, himself one of the key figures in this development, wanted to hold fast to cause and effect—being unwilling to admit that, as quoted by Monod, "God plays at dice." This attitude in what was surely one of the greatest minds of our time shows that in our animist heart of hearts we know the truth of cause and effect to be self-evident. Or, as Niels Bohr, who for many years tried to convince Einstein that God *does* play at dice, pointed out, the denaturing of common-sense concepts, such as attributing to light the properties of *both* particles *and* waves, is an *irrational* albeit scientifically useful procedure.

The covenant is coming to the end of the road because nature in its deepest aspects turns out to be incompatible with rational thought, that capital feature of man's central nervous system selected by evolution for its fitness to deal with phenomena on *existential* dimensional scales commensurate with direct human experience (give or take a few factors of ten).

It is my belief that this development—the dissolution of the covenant—presages the end of science, since there is little use in continuing to push the limits of our knowledge further and further if the results have less and less meaning for man's psyche. And instead of continuing in his efforts to analyze nature by the old rational-scientific methods, another frankly irrational and characteristically Eastern approach to fathoming nature is already gaining ground. Monod writes: "It is hard to understand . . . why [the idea that nature is objective] never cropped up in some of the loftiest civilizations, such as the Chinese, which had to learn it from the West." As a matter of fact, the Chinese knew all about the principle of objectivity when, two millenia ago, they reached the highest level of civilization, cultural as well as technological, seen until then on the face of the Earth. Already in Hellenic times, Plato's Chinese contemporary Mo Tzu had presented an elaborate argument in favor of an objective, empirical approach to natural phenomena. But once the Chinese had attained their pinnacle, they found (and for the first time in history could *afford* to find) the principle of objectivity wanting. While the Dark Ages were setting on the West, China eschewed the Moism of Mo Tzu and turned toward the Taoism of Lao Tzu, a kind of animism in reverse that projects nature into man, rather than man into nature. This turnabout changed man's ancient quest from *dominion over* to *harmony with* nature. Upon this profound change in attitude, Chinese science stagnated, while after their emergence from the Dark Ages, the Judeo-Christians of the West resumed efforts to improve their lot by dominating nature. Finally, the Western barbarians overtook China and taught it not so much the principle of objectivity as the soon-to-become-bankrupt concept of dialetical materialism.

Now it seems to be the turn of the West to learn the Taoist Natural Philosophy from the East. In the West, the ancient covenant was still intact, as long as the account of man's relation to the universe changed merely from the story of Genesis to the tale of Darwinian evolution. No, dissolution of the covenant is coming only these very days, when the attempt to fathom life is ever more being based on the sixty-four elements of the *I Ching*, the Chinese *Book of Changes*, rather than on the sixty-four elements of the genetic code, and when the notion of law and order is disappearing, not merely from our streets, but from the very universe.

* * *

Postscript (1978). Not long after the original publication of this essay, Jacques Monod presented a retrospective on his *Chance and Necessity* before the Medical Society of the World Health Organization. (The text of this retrospective was published posthumously in 1977 in the W. H. O. journal *Prospective et Santé*). Although Monod did not mention my essay specifically, I cannot help thinking that he was, in fact, addressing my critique. Or was it only a coincidence that at the outset of his reflections he expressed his surprise that in France his book ranked just behind *Love Story* as a best seller and that it also sold well in Germany, and that he noted his lack of surprise that it had only a very limited success in "Anglo-Saxon" countries? Monod surmised that *Chance and Necessity* was so successful because, throughout his book, he was trying "to pose the existential problem, not, of course, in scientific terms—since, being a metaphysical, or more exactly, a moral problem it cannot be so posed—but [merely] in terms that are not incompatible with a scientific attitude." Thus the term "metaphysical," which, in the style of positivist philosophers, Monod had used in his book only as a derogatory epithet, later turned out to be descriptive of the central aim of his project. And, retrospectively, he also recognized his principle of objectivity to be "a necessarily metaphysical postulate." How the choice of that postulate as the basis of science is to be justified Monod found "much too technical" a problem to mention in his address, but he thought that Karl Popper's criterion of falsifiability would do as the acid test of the scientific status of knowledge. As far as the meaning of "objective" is concerned, Monod admitted that there had been much confusion over the sense in which he had used that term in his book. To clear up that confusion, Monod declared that he did *not* have in mind the ordinary meaning, or antonym of "subjective," since "everybody knows that a scientist, as a subject, is not objective." What he did mean was that knowledge is "objective" provided that it rejects interpretations of phenomena in terms of final causes. This definition, based merely on the absence of a particular feature, he judged to be "very precise, purely logical, purely epistemological." Finally, as far as his "ethic of knowledge" is concerned, Monod confessed that he erred in using a term that was probably not so good, in that the relation between knowledge and ethics is actually the reverse from that implied by it. Since objective, and hence the only true, knowledge rests, according to Monod, on an ethical rejection of final causes, what is under discussion would better have been referred to as a "knowledge of ethics." Thus far from actually providing a basis for the construction of an "objective" ethical system, his solution to the "existential problem" turned out to consist merely of developing a self-consistent value system that is not in flagrant contradiction with the rejection of final causes. Evidently, by 1973 Monod had made a significant retreat from the earlier strident scientism of *Chance and Necessity*.

Contrast in Moral Traditions. Left. "Christian" Ultimate Values. ["Moses," by Rembrandt.] Right. "Pagan" Harmony. [Miroku Bosatsu (Maitreya). Camphorwood statue from the Asuka Period (552-645) at the Chugu-ji Nunnery, Nara, Japan.]

7

THE DILEMMA OF SCIENCE AND MORALS

[1974]

Ever since the sixteenth century, when Francis Bacon put forward the then novel creed that science provides a hope for a better world, there have arisen conflicts between science and morals. But right from the very start of modern science and with the case of its founder, Galileo, these conflicts were always resolved in favor of science in the long run. By the end of the nineteenth century the triumph of science over traditional, and particularly religious, morals seemed well-nigh complete. Nevertheless, there not only still arise some troublesome conflicts between science and morals but the credibility of the Baconian creed of salvation through science is itself fast losing ground in its Western heartland. This latter-day growth of anti-scientific attitudes is as serious as it is surprising because, far from its reflecting the views of ignorant rabble-rousers or religious zealots, it is occurring among the young intellectuals of the New Left. That is to say, it has infested the minds of the very group that would ordinarily furnish the recruits for the next generation of scientists. Alarmed by this development, the Old Guard has been defending the Baconian creed by means of righteous sermons. But these sermons have little effect; their language of indignant reason does not reach the ears of the

131

young infidels and does no more than preserve the courage of the true believers.

Much of the attack on science by the New Left, as well as its defense by the Old Guard, is preoccupied with the so-called misuses of science in war and in peace—with the killing and maiming of defenseless civilians, with the control and exploitation of subject peoples, and with the despoilment and pollution of the Earth by the technological fruits of modern research. The Old Guard, of course, deplores these misuses as much as the New Left. But in the view of the former it is wrong to blame science only for our problems while ignoring its contributions to our welfare. The way to avert these misuses, so the sermons usually proclaim, is not to stop doing science but to give them political and scientific remedy. Anyhow, how will we ever be able to feed the hungry of the world and to cure cancer if we turn away from science now?

In my opinion, these discussions rarely consider a deeper cause of the contemporary decline of the Baconian creed, which is philosophically more troublesome than the misuses, inasmuch as it has no remedy, even in principle. I am referring here to the moral difficulties which have arisen from some applications of science which, far from being meant to kill or enslave people or to destroy nature, are intended to augment human welfare and which nevertheless have sinister implications. It is to this latter category that some of the present and proposed applications of human biology belong. Despite their overt philanthropic intent, these applications seem monstrous and evoke the specter of Doctors Strangelove and Frankenstein. The thesis that I shall try to develop in this essay is that the moral dilemma posed by benevolent science (in contrast to its malevolent applications) is not so much that science sometimes conflicts with ethics as that the growth of scientific insights and the power that has developed from them have made it evident that the ensemble of traditional Western metaphysics and morals which spawned science in the first place is inconsistent.

* * *

According to Isaiah Berlin, the contradictory character of the Western moral tradition was discovered, or at least plainly stated, by Machiavelli a century before Galileo even opened the door to modern science. Berlin expresses the view that Machiavelli is one of the great enigmas of Western letters. For at least four centuries now, there has been a debate about just what it was that Machiavelli had intended to convey in *The Prince* and the *Discourses*, despite the fact that he was a most lucid writer. How is it that, although Machiavelli's text is perfectly clear, people continue to argue about what it is supposed to mean? Moreover, his writings have earned Machiavelli an ecumenical and everlasting hatred of men representing the whole spectrum of religious, philosophical, and political thought. How is it that his publication of a bit of advice to a Renaissance prince has managed to offend Catholics and Protestants, autocrats and democrats, reactionaries and revolutionaries across the centuries? Berlin's answer to these questions is that Machiavelli published a most disturbing insight which no ideologue who has a plan, no man who has a dream, can really accept, to wit, that the ensemble of our aims is inconsistent. Hence, the City of God cannot be realized on earth, not because of the frailties or imperfections of man but because that City is meant to satisfy mutually incompatible goals. The pope, Martin Luther, Frederick the Great, Karl Marx, and Bertrand Russell all may differ in their vision of the City of God and/or in how to go about building it, but they all share essentially the same ethical system and the fervent belief that such a thing as an ideal society can exist. No wonder that Machiavelli's subversive message that no such society is possible has made him appear as the Devil incarnate.

The contradiction to which Machiavelli drew attention is not, as has often been alleged incorrectly by commentators on *The Prince* and the *Discourses*, between morality and politics but between two incompatible systems of ethics that form part of the Western cultural heritage. One of these, which Berlin terms "Christian," envisages morality as being based on "ultimate values sought for their own sakes—values recognition of which alone enables us to speak of crimes or morally to justify and condemn anything." The other system of ethics, which Berlin terms "pagan," derives its authority from the fact that man is a social animal who lives in communities. In the pagan system there are no ultimate values, only communal purpose, and hence here moral judgments are relative rather than absolute. Or, more simply stated, the two mutually incompatible aims projected into the City of God are freedom and justice for the individual, on the one hand, and law and order for the body politic, on the other. From this insight of Machiavelli it follows, according to Berlin, "that the belief that the correct, objectively valid solution to the question of how men should live can in principle be discovered is itself, in principle, not true."

But what is the source of the belief in an objectively valid set of ethics in the first place? It is the doctrine which in one version or another has dominated Western thought since Plato "that there exists some single principle that not only regulates the course of the sun and the stars, but prescribes their proper behavior to all animate creatures." Central to this doctrine is the notion of God, or His atheistic equivalent, Eternal Reason, "whose power has endowed all things and creatures each with a specific function; these functions are elements in a single harmonious whole and are intelligible in terms of it alone . . . This unifying monistic pattern is at the very heart of traditional rationalism, religious and atheistic, metaphysical and scientific, transcendental and naturalistic, which has been characteristic of Western civilization. It is this rock, upon which Western beliefs and lives have been founded, that Machiavelli seems, in effect, to have split open."

* * *

To illustrate the ethical contradictions to which he drew attention, Machiavelli provided some concrete examples from politics, statecraft, and warfare of classical antiquity and Renaissance Italy. In this essay, I present some examples from modern science in order to try to show that Machiavelli's discovery can also illuminate its troublesome and equivocal moral role.

The first example we might consider concerns the teaching of evolution in the public schools, which evidently has come a long way from the days of the Scopes Monkey Trial in Tennessee half a century ago. In 1972 the Curriculum Commission of the California State Board of Education held hearings in response to the demand of some Christian fundamentalist groups that in the officially approved biology textbooks the biblical account of Creation ought to be presented on an equal footing with the Darwinian view as an explanation of the origin of life and of the species. Although much of the argument before the Commission pertained to the question of whether the theory of evolution is merely an unproven speculation, as alleged by the fundamentalists, or a solidly documented scientific proposition, as claimed by the biologists, the deeper point at issue was religious freedom. For the fundamentalists held that a Christian child in a tax-supported school has as much right to be protected from the dogmas of atheism as an atheist child has to be protected from prayer. Hence, it would follow that the classroom teaching of Darwinism as the only explanation of biocosmogony is an infringement of the religious freedom of Christian parents to raise their children in the faith of their choice. This argument seems completely justified, whether or not it is true as claimed in pro-Darwinian testimony at the hearings by liberal, apologist clergymen that one can be a good Christian without taking the biblical account of Genesis all that literally. After all, the fundamentalist faith *is* to take the Bible literally. But the inference that follows from admitting the justice of the fundamentalist claim is not that biology texts should give Genesis equal time with evolution. Rather, it is to be concluded that no public school system can operate effectively in a heterogeneous social setting without having its curriculum prejudice the minds of the pupils against the cherished beliefs of some of the citizens. In other words, in this case the ultimate Christian ethical aim of freedom and individual rights has to give way to the pagan aim of mounting a pedagogically effective society.

A second example is provided by recent radical criticisms directed against involuntary confinement of persons in mental hospitals and, indeed, against the very concept of insanity. For instance, Thomas S. Szasz has argued that mental illnesses are not genuine diseases and that psychiatry is not a bona fide medical specialty. One of the two main arguments put forward by Szasz in support of this proposition (the other we shall consider later) is that a medical patient can only be a person who voluntarily assumes that role and a physician can only be a person who gives treatment with the consent of his patient. Since, according to Szasz, psychiatric treatment is chiefly involuntary (overtly or covertly), insane persons are not really ill and psychiatrists are not really physicians. Psychiatric practice must, therefore, be disavowed since "in a free society, the fact that a person has an illness or that an illness be attributed to him—regardless of whether that illness is bodily or mental, literal or metaphorical—does not, and cannot, by itself justify imposing medical treatment on him against his will." Indeed, "one of our most precious rights . . . is the right to be ill—that is, the right to reject treatment, the right to die unmolested by interventions imposed on us by the state acting through its medical (or psychiatric) agencies."

Szasz's argument, like that of the fundamentalists, seems completely justified: Involuntary treatment, just as involuntary unilateral exposure to Darwinism, is incompatible with a free society. But here, too, the conclusion that follows is not that psychiatric practice ought to be disavowed but that Szasz's free society is not a workable proposition. Szasz himself seems to realize this since he would require prior consent only for the treatment of "conscious adults," thus permitting pediatrics, the treatments of which are mostly given without the patient's informed consent, to remain within the realm of legitimate medicine. Evidently, Szasz is willing to grant that in the case of children the faculty of consent is immature and that therefore others must decide the wisdom of medical treatment for them. But once having tacitly admitted that point, one is completely unreasonable to assert that there can be no abnormal persons whose chronological age and physiological state place them within the class of "conscious adults" but whose faculty of consent, for one reason or another, failed to reach maturity. Such persons, like children, are subjected to involuntary treatment simply because society looks after the welfare of those of its members who are unable to take care of themselves. Maybe Szasz is right in saying that the right to be ill and to die unmolested is one of our most precious rights, but, precious as it may be, the free exercise of that right is not possible in a functional society. Szasz is probably right, furthermore, in thinking that psychiatric practice is incompatible not only with a free but also with a just society. For while persons declared mentally ill can be subjected to involuntary psychiatric treatment without having done anyone any harm, they also can escape the normal process of criminal justice if they *have* done others great harm.

In other words, we see once more that the ultimate ethical aims of freedom and justice are in conflict with the practical social aim of communal purpose.

* * *

It is not only the Christian system of ethics which is founded on the rock that Machiavelli split open. For the monistic doctrine of an orderly universe created by God which operates by natural law and which reason can discover is also the metaphysical foundation of Western science. It is one version of Monod's "ancient covenant" between man and nature. Thus, as was recognized by Dali, but not by Crick, a Western scientist is a man who believes in God, for without this belief it would be futile to try to discover His laws. A convenient demonstration of the need for the belief in God—which the majority of contemporary scientists undoubtedly would deny, of course—was provided when Einstein affirmed his unwillingness to accept the philosophical implications of the quantum mechanical uncertainty principle in his dictum "God does not play at dice." Though Einstein was probably half-joking when he used God's name in this connection, the fact remains that it would have required a cumbersome circumlocution (such as "hidden variables") to express exactly the same sentiment without reference to God. Another demonstration of this need was provided by Crick himself in *Of Molecules and Men*, when discussing the prohibitively long computations necessary for deducing the three-dimensional conformation of proteins from the sequence of their component amino acids. Commenting on the fact that despite these cumbersome calculations proteins find their conformations all the same, Crick wrote (like Einstein, probably half-joking, but, unlike Einstein, substituting a personified "Nature" for "God" to avoid giving the impression that he is a Christian): "Nature's own analogue computer—the system itself—works so fantastically fast. Also she knows the rules more precisely than we do. But we still hope, if not to beat her at her game, at least to understand her."

Now, whereas one may reasonably doubt that Christian absolutist ethics have been more helpful than pagan relative ethics in the search for the good life, science spawned by the very same doctrine as the Christian ethics of God's lawful universe has evidently been gloriously successful. Since Galileo gave it its start, modern science has gone a long way in showing that nature is indeed accessible to reason and that, by the understanding thus obtained, man can gain extensive mastery over natural events. Thus, even though the monistic doctrine has so far received little confirmation from its application to the ethical domain, the excellent service it has rendered to modern science seems to provide for its validation. But finally, in our day, the enormous progress in science has brought to light that the doctrine of the lawful universe also embodies epistemological conflicts for science.

The epistemological contradiction that has come to light with the growth of modern physics was a major philosophical concern of Niels Bohr. He pointed out that "as the goal of science is to augment and order our experience, every analysis of the conditions of human knowledge must rest on considerations of the character and scope of our means of communication. Our basis [of communication] is, of course, the language developed for orientation in our surroundings and for the organization of human communities. However, the increase of experience has repeatedly raised questions as to the sufficiency of concepts and ideas incorporated in daily language." Accordingly, the models which modern science offers as explanations of reality are pictorial representations built on these everyday concepts. This procedure was eminently satisfactory as long as explanations were sought for phenomena that are commensurate with the events that are the subject of our everyday experience (give or take a few orders of magnitude). But this situation began to change when, at the turn of this century, physics had progressed to the stage where problems could be studied involving either tiny subatomic or immense cosmic events on scales of time, space, and mass billions of times smaller or larger than those of our direct experience. Now, according to Bohr, "there arose difficulties of orienting ourselves in a domain of experience far from that to the description of which our means of expression are adapted." For it turned out that the description of phenomena in this domain in ordinary, everyday language leads to contridictions or mutually incompatible pictures of reality. In order to resolve these contradictions, time, space, and mass had to be denatured into generalized concepts whose meaning no longer matched that provided by intuition. Eventually, it also appeared that the intuitive notion of cause and effect, central to the concept of natural law, is not a useful one for giving account of events at the atomic and subatomic level. All of these developments were the consequence of the discovery that the rational use of intuitive linguistic concepts to communicate experience actually embodies hitherto unnoticed presuppositions. And it is these presuppositions which lead to contradictions when the attempt is made to communicate events outside the experiential domain. Now, whereas the scope of science was enormously enlarged by recognizing the pitfalls of everyday language and denaturing the intuitive meaning of some of its basic concepts, a heavy price had to be paid. For, although it became possible to provide an ever more exhaustive and unified explanation of experience, that explanation came to resemble less and less the Platonic universe whose metaphysical acceptance inspired the whole enterprise of modern science in the first place. We have been duped, for if God *does* play at dice, He is not doing His job.

* * *

Though the growth of modern physics has been responsible for recognition of the deep epistemological contradictions inherent in the doctrine of the orderly universe accessible to reason, it is the growth of modern biology that has brought to light the moral contradictions inherent in the correspondent system of ethics. To appreciate the nature of these moral contradictions, we must give brief consideration to the concept that is altogether central to the Platonic ethics of which we are the heirs, namely, the *soul*. Belief in the soul has been as essential for Western morality as belief in natural law has been for Western science, the metaphysical source of both being, of course, God. The modern formulation of the problem of the soul is due to Descartes. Descartes laid the philosophical foundations for physiology (and particularly neurophysiology) by advancing the fruitful notion that the bodies of humans and animals can be regarded as machines. But since moral principles obviously do not apply to machines but do apply to humans, humans must be more than automata in human shape. The extra something that makes men more than automata is the soul, an agency that is not itself part of the body. It is from their incorporeal soul that men derive both the freedom of and the responsibility for action, without belief in which there can be no Christian ethics. For the purpose of dealing with the intersection of morals and human biology, nothing has thus far replaced the Cartesian body-soul dualism, scientistic mumbo jumbo about "objective" ethical systems based on tautological evolutionary arguments notwithstanding. (That few contemporary biologists would admit to a belief in the soul proves only that many of them resemble Molière's Monsieur Jourdain, who did not realize that he was speaking prose.)

Szasz's essay provides a handy illustration of the fact that the Cartesian dualism is very much alive today and remains the (unstated) metaphysical premise of medical ethics. Szasz's second main argument in support of the proposition that mental illnesses are not genuine diseases and psychiatrists not bona fide physicians is that insanity is not attributable to "an abnormality or malfunctioning of [the] body. . . . Strictly speaking, . . . disease and illness can affect only the body. Hence there can be no such thing as mental illness. The term 'mental illness' is a metaphor." At first sight it seems quite incredible that Szasz could claim that the abnormal behavioral symptoms associated with insanity do not derive from a malfunctioning of the body. Does he, a professor of psychiatry in the State University of New York, not know that complex aspects of human behavior are generated by an organ of the body called the brain, that the advances of neuroanatomy and neurophysiology of the past century have provided extensive insights into just how the brain manages to do its work, and that certain well-defined abnormalities or malfunctions of that organ produce psychological deficits? I imagine that Szasz does know all this, but the moral implications of that knowledge are simply unacceptable. In fact, Szasz makes plain the philosophical source of his moral rejection of psychiatric practice by accusing Freud, whom he holds (falsely) responsible for creating the metaphor "mental illness" in the first place, of a "systematic strategy for reifying and personalizing pseudomedical labels, and for stigmatizing and depersonalizing persons." Szasz evidently holds to the Platonic doctrine which informed Descartes: that the "real" person, the free and responsible agency, is not the body but the incorporeal soul. And since the soul is incorporeal, abnormalities or behavioral deficits ordinarily associated with insanity cannot be bodily ills and hence are outside the realm of medicine. Thus, to treat insane people as if they were sick is, according to Szasz, to confuse medicine with morals: "Hence, if and insofar as it is deemed that 'mental patients' endanger society, society can, and ought to, protect itself from the 'mentally ill' in the same way it does from the 'mentally healthy'—that is by means of the criminal law." Though in his polemic Szasz seems to ignore completely the insights into the workings of the human brain brought by neurology and psychology, he has nevertheless seen more clearly than many other writers the basic dilemma. And that is that the biological reification of the soul, the dissolution of the Cartesian dualism, is incompatible with the maintenance of Western ethics.

*   *   *

We may now consider the ethical conflicts surrounding two applications of human genetics. One of these is the very troublesome matter, at least for present-day American society, of the heritability of intelligence and in particular of the problem whether there exist significant racial differences in intelligence genotype. On the one hand, it seems reasonable to think that if there is a significant variation in the genetic contribution to intelligence between individuals, or between racial groups, then this factor ought to be taken into account in the organization of society. But, on the other hand, the mere acknowledgment of the existence of this factor, let alone taking it into account in social action, seems morally inadmissible, a scientistic underpinning of racist ideology. An excellent exposition of this problem was recently provided by W. Bodmer and L. L. Cavalli-Sforza, who show that the heritability of intelligence, unlike extrasensory perception and telepathy, is a genuine scientific proposition. First, it is possible to obtain a meaningful measure of intelligence through IQ tests, at least insofar as the concept of intelligence applies to the capacity to succeed in the society in whose contextual setting the tests are given. Second, there do exist significant differences in IQ between individuals and between social and racial subgroups. Third, it is possible, at least in principle, to perform studies that can ascertain the relative contribution of genetic and environmental factors to the observed differences in IQ. Bodmer and Cavalli-Sforza find that there is sufficient evidence at present to make it very likely that within a socioeconomically homogeneous group heredity does make a significant contribution to extant differences in IQ. When it comes to the considerably lower mean IQ of American blacks, however, they conclude not only that the currently available data are inadequate to ascertain whether this fact is attributable mainly to hereditary or mainly to environmental differences, but "that the question of a possible genetic basis for the race IQ differences will be almost impossible to answer satisfactorily before the environmental differences between U.S. blacks and whites have been substantially reduced. . . ." Finally, "[since] for the present at least, no good case can be made for [studies on racial IQ differences], either on scientific or practical grounds, we do not see any point in particularly encouraging the use of public funds for their support. There are many more useful biological problems for the scientist to attack."

In my opinion, this recommendation, which trivializes the problem scientifically, amounts to taking the easy way out from a serious dilemma. What if, as Bodmer and Cavalli-Sforza admit could be true, there does exist a significant genetic contribution to the mean IQ differences found between blacks and whites? They think that this "should not, in a genuinely democratic society free of race prejudice, make any difference." But if the races really differed hereditarily in intelligence, then racism would not be a "prejudice" but a true perception of the world and one of which a rational society ought to take account. For instance, in this case, the black-white disparities in socioeconomic levels would not reflect discrimination at all but merely an underlying biological reality. And hence the aim of an egalitarian, multiracial society would be just another unattainable, utopian dream. We thus encounter another Machiavellian contradiction between the two incompatible ethical systems of our heritage. The pagan ethics of communal purpose, which science serves, would demand that every effort be made to ascertain whether the member races of a multiracial society do in fact differ hereditarily in their intelligence. But the Christian ethics of ultimate values, which inspire science, holds racism to be an absolute evil in that it is subversive of the fundamental concept of the freedom and responsibility of the human soul. Hence, these ethics demand an uncompromisingly hard line against research on race intelligence. Since there must not be any hereditarily determined racial differences in intelligence, research that entertains the possibility of such differences is *a priori* evil.

The second ethically troublesome application of human genetics I shall consider concerns the purposeful manipulation of the human genotype. In a recent essay, evidently informed by the Baconian creed of scientific optimism, Bernard D. Davis provides an excellent summary overview of the practical possibilities and philosophical implications of human genetic engineering. First, Davis finds that some New Left scientists have excessively dramatized the threat posed by the possible application to the human genome of recent molecular genetic developments, mainly in order to persuade the public of the need for radical change in our government. But this exaggeration of the dangers imminent in genetic research is not likely to make the revolution; it will merely "contribute to an already distorted public view. . . . Indeed, irresponsible hyperbole on the genetic issue has already influenced the funding of research." Davis holds that, though some danger does exist from possible unwise and even malevolent applications of genetics, this danger is very small compared with the immense potential benefits. In any case, only a rather limited range of genetic manipulations, such as the repair of single-gene defects and the predetermination of sex, are realistic possibilities for the foreseeable future. By contrast, most of the more fanciful projects for the directed modification of polygenic traits, particularly those pertaining to psychological function, Davis thinks "will remain definitely in the realm of science fiction." Thus, there is little reason to wax alarmed over the imminent dangers of genetic engineering.

There is one kind of genetic manipulation, straight from the pages of science fiction, however, that Davis thinks may soon become a practical reality. This is the asexual reproduction, or cloning, of mammals, which is likely to be accomplished before long by transfer of somatic diploid nuclei from a single donor animal to enucleated eggs. Out of these eggs will grow a clone of genetically identical individuals, all possessing the genotype of the donor: "There is a considerable economic incentive to develop this procedure, since the copying of champion livestock could substantially increase food production. . . . [And] if the cloning of mammals becomes technically feasible its extension to man will undoubtedly be very tempting, on the grounds that enrichment for proved talent by this means might enormously enhance our culture, while the risk of harm seemed small."

A philosophical point of interest is that the prospect of populating the earth with clones of genetically identical humans is not, in fact, tempting at all. Why is it that, while it would be fun to have Kant, Beethoven, Bettina von Arnim, Einstein, Picasso, Clark Gable, and Marilyn Monroe living on our block, the thought of having hundreds or thousands of their replicas in town is a nightmare? Davis, too, feels apprehensive about cloning of humans; he fears that the achievements of a replica Tolstoy, Churchill, Martin Luther King, Newton, or Mozart (I drew up my own list of model genotypes before I saw Davis's) might not equal those of their isogenic prototypes. Davis thinks, furthermore, that cloning is likely to create an evolutionary danger, since the reduction in genetic diversity of the human species that would result from replacement of sexual by asexual reproduction would affect adversely its capacity to respond adaptively to sudden environmental changes. This evolutionary argument against cloning, though widely accepted by biologists, lacks logical rigor. For the very mastery over nature that would allow man to change his reproductive mode from the sexual to the asexual would presumably allow him also to make a technological (i.e., phenotypic) rather than a hereditary (i.e., genotypic) adaptive response to any putative environmental change.

No, the almost universal revulsion evoked by the prospect of cloning humans can hardly derive from practical considerations of the kind adduced by Davis. The idea of beholding a horde of look-alike human stereotypes is abhorrent even to people who are quite unaware of and who in fact lack the scientific sophistication to appreciate such arguments. The reason for the horror is, in my opinion, the belief in the uniqueness of the soul. Even though the Platonic soul is incorporeal, it is supposed to fit the body; hence, it is hard to conceive of unique souls inhabiting thousands of identical bodies. In other words, the cloned humans would not seem to be real persons but merely Cartesian automata in human shape.

That our perception of its uniqueness is, in fact, an important element in judging a being as fully human can be readily shown. For instance, the tendency of all members of a foreign race to look alike is a precondition of racism. By being thus depersonalized, the people of another race are deprived of their souls and the racist can make himself comfortable in the belief that these inferior beings are little more than animals. A similar process of depersonalization occurs in war. As is manifest in many accounts of wartime experience, soldiers can suspend the dictates of their private morality more readily in brief encounters with an unknown or even invisible enemy than they can *vis à vis* a particular member of the enemy camp (especially if he is of the same race) if an opportunity has been afforded to establish the uniqueness of his person. The faceless, homogeneous, and collective enemy has no soul; he is merely a dangerous beast outside the bounds of morality. Once recognized as a unique individual, however, the enemy acquires a soul, joins the family of man, and comes within the purview of morality. The inverse process applies to the treatment of household pets; the more the individuality of a dog or cat is recognized, the greater the tendency to personify that animal. In other words, here the perception of uniqueness causes the master to endow his pet with a soul and to raise it to the status of honorary human.

We thus encounter one more contradiction inherent in Western aims brought to light by scientific advances. The utopian dreamers of the City of God, from More to Marx, think of their perfect societies not in terms of real men but in terms of angels that embody all of the best and none of the worst human attributes. To my knowledge, diversity has never been considered an important utopian value (at least not outside the scientistic circles that try to derive values from evolutionary considerations). On the contrary, the more alike the angels are in their beauty, goodness, and intelligence, the more perfect is the vision of their society. As long as, due to the vagaries of the sexual reproductive mode, there was not the slightest chance that such angelic populations could actually arise, this seemed to be a believable dream, a hope for a better future. Only now, when advances in genetic and developmental biology have brought the asexual generation of homogeneous angelic populations within technological reach, does it suddenly become clear that this is not the kind of perfect society that we want after all. What we *do* want is the impossible: a perfect society made up of a heterogeneous collection of imperfect, unique souls, warts and all.

* * *

These conflicts and contradictions are unlikely to be resolved within the context of the Western tradition. What it would take to solve the dilemma is to abandon belief in God and His natural law and give up the righteous Christian ethical system based on absolute values and adopt instead a wholly relative system of private and social morality. That is, instead of truth and justice, wisdom and harmony would become the primary values. But is this a possible moral basis for a civilized society? It certainly is, since there already exists on Earth another great civilization, namely, the Chinese, which has this other basis. Chinese beliefs and lives are not founded on the Platonic rock that Machiavelli split open. And an examination of that other tradition shows what morality and science without God are really like. In the light of the Chinese tradition, dialectical materialism and devout Christianity can be seen to be merely minor variations on the same Platonic theme: Atheistic scientism is merely old divine wine in new bottles.

At about the time that Greek philosophers formalized the notion of the lawful universe whose mode of operation is accessible to reason, there developed in China the two complementary philosophicoethical systems of Confucianism and Taoism, which govern life there in large measure today. Confucianism is a set of down-to-earth ethical guidelines for the proper management of society. Its precepts are based on the fundamental premise that man is a social creature and that, therefore, there is virtue in harmonious social relations. These relations are made harmonious not by obedience to universally valid abstract moral principles such as freedom and justice but by exact adherence to a combination of prescribed etiquette and ritual. Taoism, on the other hand, is a transcendental, personal moral philosophy whose main relevance is for the inner life rather than for social relations. Its precepts are based on the fundamental premise that man is part of Nature and that, therefore, his life must take the path, or tao, of natural events. Man, following the tao, must abjure all striving, distrust reason, and attempt to attain a state in which he is as free from desire and sensory experiences as possible. Neither Confucianism nor Taoism invokes God (whom it does not know anyway) or Eternal Reason as the source of its authority, nor does it posit the existence of any natural law or rights of man. Rather, both systems endeavor to provide for man's harmony with his environment.

For the first few centuries of their existence Confucianism and Taoism, one advocating social engagement and the other personal withdrawal, were seen by their respective adherents as being in conflict with each other and with the Third Force of Moism, whose metaphysics, in fact, resembled the Western notion of a lawful Universe presided over by God. But then Moism disappeared and a more or less symbiotic relation between Confucianism and Taoism eventually developed. In this philosophicoethical symbiosis, the Confucian bureaucracy ran the country while the Taoist intelligentsia provided spiritual and cultural leadership. Taoism, with its focus of attention on Nature, also became the intellectual fountainhead for the development of Chinese science. But since Taoism mistrusts the powers of reason and logic and does not provide for the idea of the laws of nature, the evolution of Chinese science took a course quite different from that of Western science. Joseph Needham epitomized this difference in the following terms: "With their appreciation of the relativism and the subtlety and immensity of the universe, [the Chinese scientists] were groping after an Einsteinian world picture without having laid the foundations for a Newtonian one." Since Taoism regards the workings of Nature to be inscrutable for the theoretical intellect, Chinese science developed along mainly empirical lines. This empirical development was slow but steady, and by Renaissance times Chinese science and the technology it inspired were considerably more advanced than anything that had been achieved in the West. Indeed, much of pre-Renaissance European science fed on Chinese discoveries that had percolated from East to West. As is well known, many of the key inventions that eventually produced the transformation of medieval into modern Europe, such as gunpowder, movable type, the mechanical clock, the magnetic compass, and the stern post rudder, were of Chinese provenance. But lacking the spiritual incentive to integrate its empirical discoveries into a general theoretical framework, Chinese science remained an intellectually fragmented enterprise. Backward Western science, on the other hand, began its meteoric rise with Galileo's discovery that models built on mathematically expressible natural laws dealing with exactly measurable quantities can give a useful account of reality. Thanks to that discovery, Western science soon left Chinese science far behind. For it turned out that, contrary to the Taoist doctrine, the workings of Nature are not all that inscrutable for the intellect. Provided that the questions one asks of Nature are not too deep, satisfactory answers can usually be found. Difficulties arise only when, as I tried to show earlier in this essay, the questions become too deep and the answers that must be given to these questions are no longer fully consonant with rational thought.

A concrete example of the gulf that still separates Eastern and Western approaches to Nature and its laws was provided in testimony by Hogen Fujimoto, a representative of the Buddhist Churches of America, at the biology textbook revision hearings of the California Curriculum Commission already mentioned. Fujimoto voiced his opposition to the inclusion of the Genesis story in the school texts because this story was contrary to *his* beliefs, namely: "In the complexities of causes and subcauses one cause cannot be isolated, and is hidden within the myriads of subcauses and conditions. For this reason, the one-cause concept such as Divine Creation cannot be accepted by the Buddhists." Although Fujimoto did not seem to object to the retention in the books of Darwinian evolution, he ought to have done so. For both Bible and *The Origin of Species* are informed by the same, in the Far Eastern view, naive idea, namely, that single causes *can* be isolated and that from their isolation there evolves an explanation of the universe. Whether one thinks that God's will or that natural selection is the cause of life is, at the Eastern remove from Western doctrines, a comparatively inconsequential detail. Therefore, Buddhist children in the California schools ought to be spared exposure to the simplistic notion that the universe can be "explained" by rational thought, be it of the biblical or the Darwinian variety. Fujimoto concluded his testimony with the observation that "the question of the beginning is beyond human intellect to grasp and, therefore, should not be incorporated in the school curriculum."

In my opinion, it is highly significant that Chinese or Far Eastern philosophy is now exerting an ever-growing influence in the West. This influence is no longer confined, as it was only a few years ago, to Zen beatniks, New Left Maoists, transcendental meditation freaks, and other far-out members of the counterculture. Instead, it has reached the very pillars of society. For instance, the sudden concern among solid Establishment-type citizens for the so-called environment is a radical departure from the ancient Western aim of dominating Nature. It represents a Taoist subversion of the Baconian creed and runs counter to the quasi-religious, nineteenth-century belief in progress. It is significant in this connection that even those powerful forces whose economic interests conflict with the ecology movement, such as the petroleum and lumber industries, now feel obliged to pay lip service to the environmental cause and to claim that their unrestricted activities are needed merely for maintenance of the status quo and not, as they had claimed in the past, for progress. Similarly, the recent accommodation of the two superpowers, the United States and the Soviet Union, to end the quarter-century-long cold war is a radical departure from their traditional, righteous, reciprocal crusading fervor to smite the enemy of man. It represents a Confucian subversion of the Christian romantic ethic of the nation as the protector of the true faith and places harmony above ideological truth in international relations. This sudden change is not to be confused with a turn toward the tolerant view that "they have as much right to their opinion as we have to ours," which would still place the new situation within the context of Western ideology. Instead, the U.S.-Soviet rapprochement seems to amount to a frank acceptance of the principle that foreign policy ought to be based not on the perception of good and evil but on the goal of making a livable world.

Most of the well-meaning members of the scientific Old Guard probably welcome these two recent developments in domestic and foreign policy. But there are other epiphenomena of the turn toward the wisdom of the East that are plainly less welcome to them. Among these must be counted the declining governmental support of basic scientific research. In my opinion, this decline is attributable not so much to an ignorance by the authorities of the fact that past support of science has been a social investment with a very high return, or to the New Left propaganda about the misuses of science, as to a sincere doubt (which, according to reports by recent visitors to China, is shared by the Chinese government) of the Old Guard claim that the amelioration of the present human condition lies in the discovery of further natural laws. Instead, there seems to be a growing belief that what it will take to make the world a better place is to understand man. But whereas the notion of the laws of nature and the methods of modern science are evidently capable of giving a satisfactory account of man's physiology, his psychology does not seem to be accessible to the procedures discovered by Galileo. According to Bohr:

> The inadequacy of the mechanical concept of nature for the description of man's situation is particularly evident in the difficulties entailed in the primitive distinction between soul and body. The problems with which we are confronted here are obviously connected with the fact that the description of many aspects of human existence demands a terminology which is not immediately founded on simple physical pictures. . . . Indeed, the use of words like thought and feeling does not refer to a firmly connected causal chain, but to experiences which exclude each other because of different distinctions between the conscious content and the background which we loosely term ourselves.

This mutual exclusion is, in my opinion, at the root of the Western dilemma of science and morals.

Bibliography

Berlin, I. "The Question of Machiavelli." *New York Review of Books*, November 4, 1971, pp. 20–32.

Bodmer, W. F., and L. L. Cavalli-Sforza. "Intelligence and Race." *Scientific American*, October 1970, pp. 19–29.

Bohr, Niels. *Atomic Physics and Human Knowledge*. New York: Science Editions, 1961.

Davis, B. D. "Prospects for Genetic Intervention in Man." *Science* 170, 1279–1283 (1970).

Flew, A. "Immortality." In Paul Edwards (ed.), *Encyclopedia of Philosophy* (Vol. 4, pp. 139–150). New York: Macmillan, 1967.

Needham, Joseph. *The Grand Titration*. London: George Allen & Unwin, 1969.

Szasz, Thomas S. "Mental Disease as a Metaphor." *Nature* 242, 305–307 (1973).

III

STRUCTURALISM ON THE LAST FRONTIER

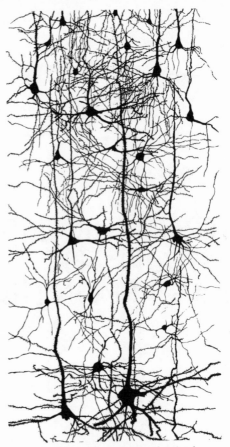

The neuronal network on the visual cortex of the human brain, as visualized under the microscope after staining the tissue with silver. The globular dark spots visible in this picture are the cell bodies of the cortical neurons that process the visual information received by the eyes. Each long and thin vertical extension emerging from a cell body is an axon, by means of which that neuron contacts other neurons in the brain. These contacts occur on the dendrites, which are seen in this picture as a mesh of thin, mainly horizontal cell extensions. [From J. L. Conel. The Postnatal Development of the Human Cerebral Cortex. Harvard University Press, Cambridge. Copyright© 1959.]

8

ABSTRACTION IN THE NERVOUS SYSTEM

[1971]

As I indicated briefly in my discussion of Monod's *Chance and Necessity*, in Chapter 6, the search for a truly "objective" understanding of nature is unlikely to succeed, because of the admittedly evolutionary rather than divine origin of the human nervous system, with which this project has to be conducted. In the following chapters, we will consider in some more detail that nervous system and the cognitive and philosophical problems posed by its intrinsic limitations.

The deepest scientific problem with which the nervous system confronts us is how it manages to work at all. Now, after the triumphs of molecular genetics have solved the puzzle of heredity, the nervous system remains the last major frontier of biological inquiry. The brain still presents us with the ancient quandary of the relation between mind and body, and it is likely that in the coming years students of the nervous system rather than geneticists will form the vanguard of biological research.

But we may ask whether scientific study of the nervous system can *ever* resolve the mind-matter paradox. Is it, in fact, likely that consciousness, the unique attribute of the brain that appears to endow its ensemble of atoms with self-awareness, will ever be explained? As was mentioned at the close of the preceding chapter, this paradox had been one of the philosophical concerns of Niels Bohr. In his address "Light and Life" before the International Congress of Light Therapy in 1932, Bohr presented what he had recognized to be the general implications of the quantum theory of atomic structure, in whose development he had played so large a part. There Bohr outlined the notion that the impossibility of describing the quantum of action, and hence what he called its "irrationality" within the purview of classical physics, is but a heuristic paradigm of how the encounter of what appears to be a deep paradox eventually leads to a higher level of understanding. "At first, this situation [i.e., the introduction of an irrational element] might appear very deplorable; but, as has often happened in the history of science, when new discoveries have revealed an essential limitation of ideas the universal applicability of which had never been disputed, we have been rewarded by getting a wider view and greater power of correlating phenomena which before might even have appeared contradictory."

These considerations, Bohr thought, would be of help when we try to fathom the nature of mind in physical terms: "The recognition of the limitation of mechanical ideas in atomic physics would much rather seem suited to conciliate the apparently contrasting points of view which mark physiology and psychology. Indeed, the necessity of considering the interaction between the measuring instruments and the object under investigation in atomic mechanics corresponds closely to the peculiar difficulties, met with in psychological analyses, which arise from the fact that the mental content is invariably altered when the attention is concentrated on any single feature of it . . . Indeed, from our point of view, the feeling of the freedom of the will must be considered as a trait peculiar to conscious life, the material parallel of which must be sought in organic functions, which permit neither a causal mechanical description nor a physical investigation sufficiently thoroughgoing for a well-defined application of the statistical law of atomic mechanics." Victor Weisskopf recently summarized Bohr's attitude in the following terms: "The awareness of personal freedom in making decisions seems a straightforward factual experience. But when we analyze the process, and follow each step in its causal connection the experience of free decision tends to disappear. . . . Bohr, an enthusiastic skier, sometimes used the following simile, which can be understood perhaps only by fellow skiers. When you try to analyze a Christiania turn in all its detailed movements, it will evanesce and become an ordinary stem turn, just as the quantum state turns into classical motion when analyzed by sharp observation." This attitude would mean nothing less than that searching for a "molecular" explanation of consciousness is a waste of time, since the physiological processes responsible for this wholly private experience will be seen to degenerate into seemingly quite ordinary, workaday reactions, no more and no less fascinating than those that occur in, say, the liver, long before the molecular level has been reached. Thus, as far as consciousness is concerned, it is possible that the quest for its physical nature is bringing us to the limits of human understanding, in that the brain may not be capable, in the last analysis, of providing an explanation of itself. Indeed, Bohr ended his 1932 lecture with the thought that "without entering into metaphysical speculations, I may perhaps add that any analysis of the very concept of an explanation would, naturally, begin and end with a renunciation as to explaining our own conscious activity."

Despite thus having to renounce the hope of ever explaining the mind, scientific studies of the nervous system have nevertheless managed to provide *some* important insights into the physical basis of mental processes. These studies can be said to have begun with Galen's definitive arguments in the second century that the brain, or cerebrum, is the site of sensation, the source of motion, and the seat of intelligence, in refutation of Aristotle's earlier claim that the heart is the locus of these functions and that the brain is merely a radiator that dissipates cardiac heat. By the seventeenth century René Descartes had formulated the problem of the generation of human behavior in terms of specific cerebral functions. Thus he asked how the brain manages to convert the data supplied to it by the sense organs into meaningful perceptions, and how it manages to issue commands to the muscles to effect appropriate action. Though Descartes supplied some speculative answers to these questions, it was only toward the middle of the nineteenth century that the first attempts were made to fathom the relation of mind and body by the methods of modern experimental science. Out of these beginnings grew the present-day discipline of neurobiology, whose avowed goal is the discovery of the anatomical, physiological, and biochemical bases of cerebral processes that underlie behaviors.

The Nervous System

The start of modern neurobiology was made possible, in part, by nineteenth-century improvements in the techniques of microscopic observation, which allowed the first insights into the cellular architecture of the brain. By the turn of this century it had been shown that the brain is a complicated network of interconnected nerve cells or *neurons*, which carry electrical signals.

Neurons are endowed with two singular features that make them particularly suitable for this purpose. First, unlike most other cell types, they possess long and thin extensions: axons. With their axons neurons reach and come into contact with other neurons at distant sites and thereby form an interconnected network extending over the entire animal body. Second, unlike most other cell types, neurons give rise to electrical signals in response to physical or chemical stimuli. They conduct these signals along their axons and transmit them to other neurons with which they are in contact. The interconnected network of neurons and its traffic of electrical signals forms the nervous system.

Like Roman Gaul, the nervous system is divisible into three parts: (1) an input, or *sensory*, part that informs the animal about its condition with respect to the state of its external and internal environment; (2) an output, or *effector*, part that produces motion by commanding muscle contraction, and (3) an *internuncial* part (from the Latin *nuncius*, meaning messenger) that connects the sensory and effector parts. The most elaborate portion of the internuncial part, concentrated in the head of those animals that have heads, is the brain.

The processing of data by the internuncial part consists in the main in making an *abstraction* of the vast amount of data continuously gathered by the sensory part. This abstraction is the result of a selective destruction of portions of the input data in order to transform these data into manageable categories that are meaningful to the animal. It should be noted that the particular command pattern issued to the muscles by the internuncial part depends not only on here-and-now sensory inputs but also on the history of past inputs. Stated more plainly, neurons can learn from experience. Until not so long ago attempts to fathom how the nervous system actually manages to abstract sensory data and learn from experience were confined mainly to philosophical speculations, psychological formalisms or biochemical naïvetés. In recent years neurophysiologists, however, have made some important experimental findings that have provided for a beginning of a scientific approach to these deep problems. Here I can do no more than describe briefly one example of these recent advances and sketch some of the insights to which it has led.

Before discussing these advances we must give brief consideration to how electrical signals arise and travel in the nervous system. Neurons, like nearly all other cells, maintain a difference in electric potential of about a tenth of a volt across their cell membranes. This potential difference arises from the unequal distribution of the three most abundant inorganic ions of living tissue, sodium (Na^+), potassium (K^+) and chloride (Cl^-), between the inside of the cell and the outside, and from the low and unequal specific permeability of the cell membrane to the diffusion of these ions. In response to physical or chemical stimulation the cell membrane of a neuron may increase or decrease one or another of these specific ion permeabilities, which usually results in a shift in the electric potential across the membrane. One of the most important of these changes in ion permeability is responsible for the action potential, or *nerve impulse*. Here there is a rather large transient change in the membrane potential lasting for only one or two thousandths of a second once a prior shift in the potential has exceeded a certain much lower threshold value. Thanks mainly to its capacity for generating such impulses, the neuron (a very poor conductor of electric current compared with an insulated copper wire) can carry electrical signals throughout the body of an animal whose dimensions are of the order of inches or feet. The transient change in membrane potential set off by the impulse is propagated with undiminished intensity along the thin axons. Thus the basic element of signaling in the nervous system is the nerve impulse, and the information transmitted by an axon is encoded in the frequency with which impulses propagate along it.

Neurophysiologists have developed methods by which it is possible to listen to the impulse traffic in a single neuron of the nervous system. For this purpose in recording electrode with a very fine tip (less than a ten-thousandth of an inch in diameter) is inserted into the nervous tissue and brought very close to the surface of a neuron. A neutral electrode is placed at a remote site on the animal's body. Each impulse that arises in the neuron then gives rise to a transient difference in potential between the recording electrode and the neutral electrode. With suitable electronic hardware this transient potential difference can be displayed as a blip on an oscilloscope screen or made audible as a click in a loudspeaker.

The point at which two neurons come into functional contact is called a *synapse*. Here the impulse signals arriving at the axon terminal of the *presynaptic* neuron are transferred to the *postsynaptic* neuron that is to receive them. The transfer is mediated not by direct electrical conduction but by the diffusion of a chemical molecule, the transmitter, across the narrow gap that separates the presynaptic axon terminal from the membrane of the postsynaptic cell. That is to say, the arrival of each impulse at the presynaptic axon terminal causes the release there of a small quantity of transmitter, which reaches the postsynaptic membrane and induces a transient change in its ion permeability. Depending on the chemical identity of the transmitter and the nature of its interaction with the postsynaptic membrane, the permeability change may have one of two diametrically opposite results. On the one hand it may increase the chance that there will arise an impulse in the postsynaptic cell. In that case the synapse is said to be excitatory. On the other hand it may reduce that chance, in which case the synapse is said to be inhibitory. Most neurons of the internuncial part receive synaptic contacts from not just one but many different presynaptic neurons, some axon terminals providing excitatory inputs and others inhibitory ones. Hence the frequency with which impulses arise in any postsynaptic neuron reflects an ongoing process of summation, more exactly a temporal integration, of the ensemble of its synaptic inputs.

The Visual Pathway

We are now ready to proceed to our example of an important advance in the understanding of the internuncial nervous system, the analysis of the visual pathway in the brain of higher mammals. It is along this pathway that the visual image formed on the retina by light rays entering the eye is transformed into a visual percept, on the basis of which appropriate commands to the muscles are issued. The visual pathway begins at the mosaic of approximately 100 million primary light-receptor cells of the retina. They transform the light image into a spatial pattern of electrical signals, much as a television camera does. Still within the retina, however, the axons of the primary light-receptor cells make synapses with neurons already belonging to the internuncial part of the nervous system. After one or two further synaptic transfers within the retina the signals emanating from the primary light-receptor cells eventually converge on about a million retinal ganglion cells. These ganglion cells send their axons into the optic nerve, which connects the eye with the brain. Thus it is as impulse traffic in ganglion-cell axons that the visual input leaves the eye.

In 1953 Stephen W. Kuffler, who was then working at Johns Hopkins University, discovered that what the impulse traffic in ganglion-cell axons carries to the brain is not raw sensory data but an abstracted version of the primary visual input. This discovery emerged from Kuffler's efforts to ascertain the ganglion-cell *receptive field*, or that territory of the retinal receptor-cell mosaic whose interaction with incident light influences the impulse activity of individual ganglion cells. For this purpose Kuffler inserted a recording electrode into the immediate vicinity of a ganglion cell in a cat's retina. At the very outset of the study Kuffler made a somewhat unexpected finding, namely that even in the dark, retinal ganglion cells produce impulses at a fairly steady rate (20 to 30 times per second) and that illuminating the entire retina with diffuse light does not have any dramatic effect on that impulse rate. This finding suggested paradoxically that light does not affect the output activity of the retina. Kuffler then, however, projected a tiny spot of light into the cat's eye and moved the image of the spot over various areas of the retina. In this way he found that the impulse activity of an individual ganglion cell does change when the light spot falls on a small circular territory surrounding the retinal position of the ganglion cell. That territory is the receptive field of the cell.

On mapping the receptive fields of many individual ganglion cells Kuffler discovered that every field can be subdivided into two concentric regions: an "on" region, in which incident light increases the impulse rate of the ganglion cell, and an "off" region, in which incident light decreases the impulse rate. Furthermore, Kuffler found that the structure of the receptive fields divides retinal ganglion cells into two classes, on-center cells, whose receptive field consists of a circular central "on" region and a surrounding circular "off" region, and off-center cells, whose receptive field consists of a circular central "off" region and a surrounding circular "on" region. In both on-center and off-center cells the net impulse activity arising from partial illumination of the receptive field is the result of an algebraic summation; two spots shining on different points of the "on" region cause a more vigorous response than either spot alone, whereas one spot shining on the "on" and the other on the "off" region give rise to a weaker response than either spot alone. Uniform illumination of the entire receptive field, the condition that exists under diffuse illumination of the retina, gives rise to virtually no response because of the mutual cancellation of the antagonistic responses from "on" and "off" regions.

It could be concluded, therefore, that the function of retinal ganglion cells is not so much to report to the brain the intensity of light registered by the primary receptor cells of a particular territory of the retina as it is to report the degree of light and dark contrast that exists between the two concentric regions of its receptive field. As can be readily appreciated, such contrast information is essential for the recognition of shapes and forms in the animal's visual field, which is what the eyes are mainly for. Thus we encounter the first example in our discussion of how the nervous system abstracts by selective destruction of information. The light intensity data gathered by the primary light receptor cells are selectively destroyed in the algebraic summation process of "on" and "off" responses and thereby transformed into the perceptually more meaningful light and dark contrast data.

When one thinks about the neuronal circuits that might be responsible for this retinal abstraction process the first possibility that comes to mind is that they embody the antagonistic function of excitatory and inhibitory synaptic inputs to the same postsynaptic neuron. Thus one might have supposed that to produce an on-center receptive field the axon terminals of primary receptor cells from the central "on" territory simply make excitatory synapses and those from the peripheral "off" territory make inhibitory synapses upon their retinal ganglion cell. Detailed analyses of the anatomy and physiology of retinal neurons carried out in recent years have shown that, on the one hand, the real situation is very much more complicated than this simplest of pictures, but that, on the other hand, the actual neuronal circuits do involve matching of excitatory and inhibitory synapses in the pathways leading from primary light receptors of antagonistic receptive field regions to the ganglion cell.

In the late 1950s David Hubel and Torsten Wiesel began to extend these studies on the structure and character of visual receptive fields to the next higher stage of information processing. For this purpose they examined the further fate of the impulse signals carried by the million or so retinal ganglion cell axons in the optic nerve from the eye to the brain. After a way-station in the forebrain, which for the purpose of this discussion can be considered a simple one-for-one impulse relay, the signal output of the retinal ganglion cells reaches a particular area of the cerebral cortex at the lower back of the head, designated as the *visual cortex*. Here the incoming axons make synaptic contacts with neurons of the cortex. The first cortical neurons contacted by the axon projecting from the eye in turn send out their axons to other cells in the visual cortex for further processing of the visual input. But onwards from there one must yet find the trail that eventually leads to those other areas of the brain, where, if the sensory percept is to elicit a behavioral act, commands must be issued to the muscles.

Hubel and Wiesel observed the impulse activity of individual neurons of the visual cortex in response to various light stimuli projected on a screen in front of the eyes and found that also these cortical neurons of the visual pathway respond only to stimuli falling on a limited retinal territory of light receptor cells. But the character of the receptive fields of cortical neurons turned out to be dramatically different from that of the retinal ganglion cells. Instead of having circular receptive fields with concentric "on" and "off" regions, the cortical neurons were found to respond to straight-line edges of light-dark contrast, such as bars of light on a dark background. Furthermore, to produce its optimum response, the straight-line edge must be in a particular orientation in the receptive field. Thus a light bar projected vertically on the screen that produces a vigorous response in a particular cortical cell will no longer elicit the response as soon as its projection is tilted slightly away from the vertical. In their first studies, Hubel and Wiesel found two different classes of cells in the visual cortex: simple cells and complex cells. The response of simple cells demands that the straight-edge stimuli must not only have a given orientation but also a precise position in the receptive field. The stimulus requirements of complex cells are less demanding, however, in that their response is sustained upon parallel displacements (but not upon tilts) of the straight-edge stimuli within the receptive field. Thus the process of abstraction of the visual input begun in the retina is carried to higher levels in the visual cortex. The simple cells, which are evidently the very next abstraction stage, transform the data supplied by the retinal ganglion cells concerning the light-dark contrast at individual points of the visual field into information concerning the contrast present at particular straight-line sets of points. This transformation is achieved by selective destruction of the information concerning just how much contrast exists at just which point of the straight-line set. The complex cells carry out the next stage of abstraction. They transform the contrast data concerning particular straight-line sets of visual-field points into information concerning the contrast present at parallel sets of straight-line point sets. In other words, here there is a selective destruction of the information concerning just how much contrast exists at each member of a set of parallel straight lines.

The neuronal circuits responsible for these next stages of abstraction of the visual input can now be fathomed. Let us consider first the simple cell of the visual cortex that responds best to a light bar in a dark background projected in a particular orientation and position on the retinal receptor-cell mosaic. That simple cell is so connected to the output of the retina that it receives synaptic inputs from axons reporting the impulse activity of a set of on-center retinal ganglion cells with receptive fields arranged in a straight line. Thus a light bar falling on all of the central "on" but on none of the peripheral "off" regions of this row of receptive fields will activate the entire set of retinal ganglion cells and provide maximal excitation for the simple cortical cell. If the retinal projection of the bar is slightly displaced or tilted, however, some light also will strike the peripheral "off" regions and the excitation provided for the simple cell is diminished.

We next consider the complex cortical cell that responds to a light bar of a particular orientation in any one of several parallel positions in the receptive field. This response can be easily explained on the basis that the complex cell receives its synaptic inputs from the axons of a set of simple cortical cells. All the simple cells of this set would have receptive fields that respond optimally to a light bar projected in the same field orientation, but they differ in the field position of their optimal response. A suitably oriented light bar projected anywhere in the complex receptive field will always activate one of the component simple cells and so also the complex cell.

In their later work Hubel and Wiesel were able to identify cells in the visual cortex whose optimal stimuli reflect even higher levels of abstraction than parallel straight lines, such as straight-line ends and corners. But it is not so clear at present how far this process of abstraction by convergence of communication channels can be imagined as going. In particular, should one think that there exists for every pattern of whose specific recognition an animal is capable at least one particular cell in the vertebrate cortex which responds with impulse activity when that pattern appears in the visual field? We will reconsider this question in a later chapter.

In any case these neurobiological insights into the visual pathway show that information about the world reaches the mind not as raw data but as highly abstract structures which are the result of a preconscious set of step-by-step transformations of the sensory input. Each transformation step involves the selective destruction of information according to a program which pre-exists in the brain. Under this program our visual preception of the world is filtered through a stage in which the input is processed in terms of straight lines, because of the manner in which the input channels coming from the primary light receptors of the retina are connected to the brain. This fact cannot but have profound psychological consequences; evidently a geometry based on straight, parallel lines, and hence by extension, on plane surfaces, is most immediately compatible with our mental equipment. This need not have been this way, since—at least from the neurophysiological point of view—the retinal ganglion cells might just as well have been connected to the higher cells in the visual cortex in such a way that their concentric "on" and "off" center receptive fields form arcs rather than straight lines. If evolution had given rise to that other circuitry, curved rather than plane surfaces would have been our primary spatial concept. Hence neurobiology has now shown what philosophical speculation led Immanuel Kant to claim 200 years ago: Euclidean geometry and its nonintersecting coplanar parallel lines is the "natural" geometry, at least for man. Non-Euclidean geometries of convex or concave surfaces, though our brain is evidently capable of conceiving them, are more alien to our built-in spatial perception processes. Apparently, a beginning has now been made in the providing of a scientific account of the relation between reality and the mind. But just how much further we can expect to progress beyond that beginning will be considered in a later chapter.

Bibliography

Kuffler, S. W. "Discharge Patterns and Functional Organization of the Mammalian Retina." *J. Neurophysiol.* 16, 37–68 (1953).

Kuffler, S. W., and J. G. Nicholls. *From Neuron to Brain.* Sunderland, Mass.: Sinauer Assoc., 1976 [This text contains a fuller, generally comprehensible treatment of the subject matter covered in this chapter.]

Hubel, D. H., and T. N. Wiesel. "Receptive Fields, Binocular Interaction and Functional Architecture in the Cat's Visual Cortex." *J. Physiol.* 160, 106–154 (1962).

Hubel, D. H., and T. N. Wiesel. "Receptive Fields and Functional Architecture in Two Non-striate Visual Areas (18 and 19) of the Cat." *J. Neurophysiol.* 28, 229–289 (1965).

Siamese Cat. The cascade of physiological, morphological, neurological, and behavioral effects engendered by the temperature-sensitive character of the product of one of its genes demonstrates the difficulty of fathoming the meaning implicit in the genetic information. [Photograph by Walter Chandoha.]

9

GENES
AND THE
EMBRYO

[1975]

The neurobiological studies summarized in the preceding chapter have sought to account for mental processes in terms of the structure and function of the adult nervous system. But there is also another deeply mysterious biological aspect of the mind-body problem, namely the genesis of the cerebral apparatus during the development of each individual from fertilized egg to adult, that is to say, during his ontogeny. How, one may ask, is it possible for the neuronal circuits to which mental processes are attributable, the precisely interconnected cellular components that make up the nervous system, to arise in the first place? That this question still awaits its answer is not surprising, inasmuch as the ontogenetic mechanisms by which any part of a multi-cellular organism comes into being still remain shrouded in mystery.

Until the latter part of the eighteenth century the dominant view of the nature of ontogeny had been that of "preformation." The preformationists envisaged that the fertilized egg already contains an invisible, miniaturized version of the adult, a homunculus, and that ontogeny consists simply in the increase in size of the homunculus from microscopic to macroscopic dimensions. Hence this view necessarily led to the belief that all later generations of the human race had already been pre-formed, one into the other, in—depending on the relative roles assigned to male and female in this infinite recessional system of Chinese boxes—Adam or Eve. To escape this absurd conclusion Caspar F. Wolff put forward the alternative view of "epigenesis." Wolff thought that the fertilized egg, far from harboring any homunculus, contains no organized structures at all, being composed of an undifferentiated protoplasm. Accordingly, the epigenetic development by which each embryo arises *de novo* from the egg represents a process which consists not only of growth but also of morphogenesis and differentiation of the living substance. When, upon the advent of the cell theory in the nineteenth century, it came to be understood that the em-bryo is the product of a series of successive cell divisions, start-ing from the single cell of the egg, the morphogenesis and dif-ferentiation called for by the theory of epigenesis was seen to pertain, not to the living substance as a whole, but to individual cells: although the billions of cells that make up an adult or-ganism are all direct descendants of the same ancestral cell, the various members of that cell colony have very different proper-ties and perform very different functions.

* * *

The first coherent scheme of embryonic cell differentiation was put forward by August Weismann in the 1880s. Weismann proposed that differentiation of body cells arises from an un-equal partition of the hereditary substance in successive cell divisions. That is to say, he envisaged that cell differentiation is the consequence of differentiation of the cell nucleus, arising from a selective loss of what we now call the parental genes. According to Weismann the total gene complement, or genome, would be preserved intact only during cell divisions in the germ line, so that the germ cells—eggs and sperm—carried by the sexually mature adult can pass on to the offspring the complete parental genome. Weismann's scheme fell into disfavor during the early years of this century, although no really critical evi-dence had been adduced which proved that a covert genetic differentiation of the cell nucleus is *not* responsible for the overt differentiation of the cell. At last, in the 1960s, an experimental disproof of Weismann's theory became possible when techniques were developed for transplanting the nucleus of a differentiated adult cell to an egg whose own nucleus had been removed. Thanks to this technique, J. B. Gurdon could show that a tadpole can develop from a frog's egg into which the nucleus obtained from a differentiated cell of a tadpole intestine had been transplanted. Thus Gurdon's experiment finally disposed of the nuclear differentiation theory, in that it showed that the nucleus of a differentiated intestinal cell still carries all the genes necessary for instructing the frog egg how to produce a frog. It was thus concluded that cell differentiation is not the consequence of a permanent change in character of the cell genome but must, instead, derive from a differential expression of the myriad of genes embodied by that genome. Hence it came to be believed that the explanation of cell differentiation must be sought in terms of the regulation of gene function.

Gurdon's discovery made it appear that the time was now ripe for molecular geneticists to find the solution to the puzzle of embryonic development. In particular, it seemed plausible to think that the regulation of gene function is effected by special control genes and that, therefore, the concept of genetic information, which proved of such enormous utility in the growth of molecular genetics, ought also to be of help for fathoming the genesis of the embryo. I myself shared that general belief in the late 1960s and thought that the specific character of differentiated cells appears to derive from the turn-on and turn-off of various genes by a regulatory mechanism not unlike the "operon" proposed for bacteria by Monod and his colleague François Jacob. Although I admitted the possibility that higher organisms might employ also regulatory circuits other than the operon, I believed that by simple extensions of molecular-genetic lore one can already imagine what these circuits are likely to be. As it turned out, my confidence in the molecular genetic approach to embryology was not warranted, since that approach has made very little headway in the intervening decade. Although a tremendous amount of relevant data has been accumulated meanwhile, few theoretical insights have been reached. In searching for an explanation for this, for molecular geneticists quite unexpected lack of progress, I have come to realize that it may be attributable to a fundamental difficulty of extending the concept of genetic information from an individual gene to the entire gene complement of an organism, or its *genome*.

* * *

As was spelled out thirty years ago by Erwin Schrödinger in his seminal book *What Is Life?* the gene can be viewed as an information carrier whose physical structure corresponds to an aperiodic succession of a small number of isomeric elements of a hereditary codescript. Eventually, during the Dogmatic Period of molecular genetics, the gene was identified as a segment of a double helical DNA molecule residing in the chromosomes of the cell nucleus. The isomeric elements of the hereditary code-script turned out to be the four nucleotide bases adenine, guanine, thymine, and cytosine, which embody the genetic information by their aperiodic linear sequence along the DNA molecule. And as far as the meaning of the information contained in an individual gene is concerned, it was first put forward as an *a priori* dogmatic postulate and later established as an empirical fact, that the linear sequence of DNA nucleotide bases stands for the linear sequence of amino acids of a particular protein molecule. The two sequences are related to each other via a genetic code, under which each nucleotide base triplet stands for one of the twenty kinds of amino acids of which protein molecules are built. However, the chromosomal DNA also contains some nucleotide base sequence segments which do not stand for protein molecules, and, strictly speaking, do not constitute genes. Some of these segments serve as templates for the assembly of nucleic acid components of the cellular apparatus for protein synthesis, such as transfer and ribosomal RNA molecules, and other segments serve as control sites at which the expression of the genes is regulated. Thus it is fair to say that there exists at present a highly satisfactory state of understanding of the nature of the informational content and meaning of the structural elements of the genetic material. However, the present status of the problem of the overall meaning of the whole genome is not really all that satisfactory.

What, in fact, *is* the overall meaning of the information embedded in the genome? What does it represent? A not uncommon answer to this question is that the DNA molecules of the genome obviously amount to a one-dimensional representation of the whole organism. For instance, the astronomer Carl Sagan suggested at a Conference on Communication With Extraterrestrial Intelligence that to transmit via radio signals the entire DNA nucleotide base sequence of the cat genome to a Distant Alien Civilization would be equivalent to sending the Aliens the cat itself. Sagan's suggestion, though made partly in jest, allows us to recognize that the correct answer to the question into the meaning of the genome is not so obvious. On the contrary, what *is* obvious is that the Alien Intelligence, even if it possessed the table of the terrestrial genetic code, would not be able to reconstruct the cat from its DNA nucleotide base sequence. To make this reconstruction, the Aliens would have to know a good deal more about terrestrial life than the formal relations between DNA nucleotide base sequences and protein amino acid sequences. What they would have to know, above all, is that the cat, or the feline *phenome,* arises by an epigenetic process from a fertilized egg containing the feline DNA nucleotide sequences. Moreover, the Aliens would have to understand the nature of the epigenetic relation between phenome and genome, an understanding which we unfortunately still lack.

Why, then, do we still lack this understanding? Why did it turn out to be so difficult to extend the great insights into the nature of gene structure and function provided by molecular genetics to the deep problem of embryonal development? As C. H. Waddington set forth well before the informational theories of molecular genetics had even received their experimental validation, the genetic information does not represent an organism but merely some functional components of an *epigenetic landscape.* What Waddington meant by this poetic term is a plot of multivariant functional relations in multidimensional space. In this space, developmental time is the independent variable and the properties that describe both the organism and its environment are the dependent variables. The functional relations from which this landscape is constructed are the chemical and physical processes which relate the changes in these properties with the flow of ontogenetic time. The topography of this landscape, therefore, represents the developmental pathways along which the embryo moves from fertilized egg to adult. Waddington's main reason for using the landscape metaphor was to point out that in this space the developmental pathways are bound to form a system of interconnected valleys that slope "downward" from the summit of the egg in the direction of ontogenetic time towards the "sea level" of the adult organism. This feature would assure that the pathways are relatively resistant to perturbations of the functional relations and to fluctuations in the dependent variables, and hence guarantee a reasonably invariant relation between genome and phenome. The role of the genes in shaping this landscape derives from their control of critical chemical processes in the developmental sequence, or (as we now know) from their governance of the production of protein molecules capable of catalyzing specific chemical reactions.

The idea of the epigenetic landscape certainly brings us closer to an understanding of the relation between the genetic infor-

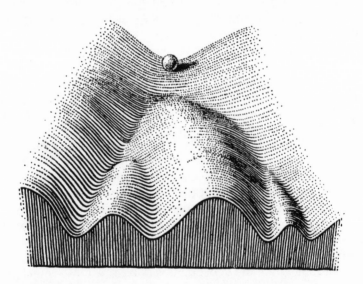

Waddington's "Epigenetic Landscape." In this picture, developmental time is running toward the viewer, and developmental potential is represented by the downward direction. The ball indicates the embryo, which may roll downward toward the sea level of the adult organism along one of several alternative paths. [From C. H. Waddington, The Strategy of the Genes. *George Allen & Unwin, Ltd., London, 1957.]*

mation and the organism to which it gives rise. And it can fairly be said that the discovery of the functional relations of that landscape, or as François Jacob has called them, "the algorithms of the living world," ought to be one of the main goals, if not *the* main goal, of contemporary embryology. But I give it as my opinion that, thus far at least, this goal has not generally been brought into sufficiently sharp conceptual focus, because of semantic difficulties inherent in the notion of "meaning," as applied to the genetic information.

* * *

It is not without irony that the problem of meaning is no less troublesome in the domain of human communication, for which the scientific concept of information was developed in the first place, than in the domain of genetics to which semantic notions have been extended by molecular geneticists. Since the question of how meaning arises from language, or even what it is we are saying about a word when we say what it means, still awaits its answer, it is hardly surprising that we encounter also conceptual difficulties with the metaphorical use of semantic terminology in genetic biology. It so happens, however, that some of the philosophical contributions to the problem of linguistic meaning can assist us also in the troublesome matter of the meaning of the genetic information.

One such philosophical contribution to semantics that can help us to understand how meaning arises from the genetic information is the recognition that the meaning of semantic structures may depend on the *context* in which they are produced. For instance, linguists draw attention to the fact that although the *literal* meaning of the sentence "I want you to shut the door" is a declaration of the speaker's state of mind, in the context of polite social intercourse the usually intended meaning of that sentence is, in fact, a command. A related, context-dependent distinction between two different kinds of meaning, which I believe is useful for our present purpose, is that between *explicit* and the *implicit* meaning. The explicit meaning is that which a semantic structure has by virtue of the syntactic relation of its elements. Hence, the explicit meaning can be extracted from the structure by subjecting it to a linguistic analysis. The implicit meaning, by contrast, is not really contained in the structure itself and arises secondarily from the explicit meaning by virtue of its context. For instance, the explicit meaning of the sentence "John Smith is traveling to New York" is that a particular individual is on his way to a particular geographical location. However, depending on the context in which the sentence is produced, it can also have a large number and variety of implicit meanings. For instance, if it is produced at San Francisco airport, it would imply that Mr. Smith is about to board a particular flight, that his suitcase is waiting to be loaded on a particular baggage truck, that he cannot be reached by telephone for the next six hours, and so on. The same sentence with the same explicit meaning would carry a different set of implicit meanings if it were produced at a roadside service station in Colorado.

Actually, it is difficult to draw a sharp line of demarcation between explicit and implicit meanings, since the extraction of the explicit meaning is itself rarely context-free. For instance, in the example just given, the intrinsic ambiguity in the explicit meaning of whether "New York" refers to the state or the city is resolved by the San Francisco airport context under which the term can be safely assumed to refer to the city. Thus the distinction between explicit and implicit meaning is relative rather than absolute, with a meaning being the less explicit and the more implicit the more dependent it is on the context. Furthermore, because of its high degree of dependence on the context, the implicit meaning is open-ended, in that it can become ever more remote from the explicit meaning as the context is widened.

When we apply this distinction to the semantic relation between genome and phenome it becomes evident that the explicit meaning of the genetic information consists of the protein amino acid sequences encoded in the genes, and of the nucleotide base sequences of the ribosomal and transfer RNA molecules encoded in other non-genic DNA sectors. The explicit meaning would include also those physico-chemical properties of the non-genic control DNA segments, such as operators that serve regulatory functions. These meanings are explicit in the sense that they can be extracted from an analysis of the DNA nucleotide base sequence itself, provided one knows that the DNA base sequence is transcribed into a complementary RNA base sequence, and one has access to the table of the genetic code. But these explicit meanings form only the basic skeleton of the functional relations that shape the epigenetic landscape. The bulk of these relations are merely implicit in the genetic information.

By way of an example of such an implicit meaning we may consider the three-dimensional conformation of protein molecules. Although it is in some sense true that the spatial conformation of a protein molecule is "genetically determined", this "determination" devolves from the DNA-encoded specification of the one-dimensional amino acid sequence of the molecule. Once assembled from its constituent amino acids, the protein molecule automatically folds to assume its functional three-dimensional structure. The physico-chemical principles which govern this folding are in part understood, although it is not yet possible (but soon may be) to predict the three-dimensional structure which a protein with a given amino acid sequence will assume. But it is important to note that these folding rules are nowhere represented in the DNA nucleotide base sequences, being part of the context rather than of the genetic informational structures. The enzymatic function of the protein molecule, which can also be said to be "genetically determined," is an even more purely implicit meaning of the genetic information than the spatial conformation. Once a protein molecule *has* been assembled specifically from its constituent amino acids and *has* assumed its specific three-dimensional structure, certain parts of that structure turn out to possess the power to catalyze some particular chemical reaction. The stereochemical principles which govern that catalysis are also partly understood, although (to my knowledge, at least) it is not yet possible to predict on the basis of the known three-dimensional structure of a protein molecule the kind of reaction which that molecule can catalyze. It goes without saying that the principles of chemical catalysis are not represented in the DNA nucleotide base sequences either; they enter into the meaning of the genetic information at a second-order level of a contextual hierarchy, at whose first-order level we found the protein folding process.

This procedure of identifying implicit meanings of the genetic information can be continued almost indefinitely to higher and higher levels of the contextual hierarchy. For instance, the physiological function of a chemical substance whose formation is catalyzed by a particular protein molecule can likewise be said to be "genetically determined", as can be the overt behavioral feature to which that physiological function gives rise. The nearly unlimited horizon of implicit meanings of the genetic information shows that, as has long been recognized, the notion of the "inborn nature," or genetic determination of characters is so all inclusive as to be nearly devoid of meaning. After all, there is no aspect of the phenome to whose determination the genes cannot be said to have made their contribution. Thus it transpires that the concept of genetic information, which in the heyday of molecular biology was of such great heuristic value for unraveling the structure and function of the genes, i.e., the explicit meaning of that information, is no longer so useful in this later period when the epigenetic relations which remain in want of explanation represent mainly the implicit meaning of that information.

* * *

In order to demonstrate the pertinence of these abstract semantic discussions for present-day biology, we may consider one presently particularly active area of research, namely the study of the development of the metazoan nervous system. The nervous system is an especially suitable object for epigenetic investigations, because, as neuroanatomical and neurophysiological studies have shown, it is the precise manner of the interconnection of the cellular components of that system to which the organism's behavior is attributable. Thus here an unambiguous definition of a phenome in want of explanation can be provided in terms of a circuit diagram of specified cellular elements. Although we cannot be sure as yet that it *will* be possible to give an account of the epigenetic landscape that produces the neural circuitry, we can be reasonably confident that the explanation, if found, will not be trivial.

The general problem of the development of the nervous system has been formulated by Seymour Benzer, one of the veteran molecular geneticists who, with the advent of the Academic Period, turned his attention to the nervous system, in the following terms:

When the individual organism develops from a fertilized egg, the one-dimensional information arrayed in the linear sequence of the genes on the chromosomes controls the formation of a two-dimensional cell layer that folds to give rise to a precise three-dimensional arrangement of sense organs, central nervous system and muscles. These elements interact to produce the organism's behavior, a phenomenon whose description requires four dimensions at least. Surely the genes, which so largely determine anatomical and biochemical characteristics, must also interact with the environment to determine behavior. But how?

One possible answer to the question of how the genes determine behavior is that they, in fact, contain the information for the circuit diagram of the nervous system. However, it has been argued that the circuitry of the nervous system cannot, in fact, be genetically determined because the total amount of genetic information does not suffice to specify the neuronal connections that need to be made. According to this argument, the linear sequence of roughly 10^{10} DNA nucleotide bases in the genome of a higher vertebrate animal contains an upper limit of 2×10^{10} bits of information (since each base, being one of four possible types, embodies 2 bits). On the other hand, if each one of the roughly 10^{10} cells in the nervous system of such an animal were connected to just two other cells, then it would require of the order of $10^{10} \log_2 10^{10}$, or 3×10^{11} bits to specify this network. Thus even under the most absurdly oversimplified view of the complexity of the nervous system, the total information content of the genetic material, even if it had no nonnervous determinative role, would be too low by an order of magnitude to allow the specification of nerve cell connections.

Although this antigenetic argument has little merit, it is useful to examine it because it exemplifies two not uncommon errors of thought which must be corrected before the relation of the genome to the development of the nervous system phenome can be profitably considered. The first of these errors derives from a spurious application of information theory to biological problems. That is to say, it derives from the failure to recognize that the quantitative concept of information applies only to processes in which the probabilities of realization of alternative outcomes are known or clearly defined. To illustrate this point, we may consider a biological example which bears some formal analogies to the problem of the circuitry of the nervous system but which is presently much better understood, namely the determination of the structure of protein molecules. A typical species of protein molecule consists of about 300 amino acid building blocks, or of about 4000 atoms held to each other in specific chemical linkage, with each atom, on the average, being connected to about two other atoms. We may ask how many bits of information are needed to specify the chemical structure of that protein molecule. If we were to proceed by the same calculation as that which was just applied to the nervous system, we would reckon that about $4 \times 10^3 \log_2 (4 \times 10^3)$ or 5×10^4 bits are needed. But here we encounter an apparent paradox, because the gene that encodes the chemical structure of a 300-amino acid protein molecule consists only of a sequence of about 900 nucleotide bases, and hence contains a maximum of $2 \times 900 = 1800$ bits of information. Thus the information content of the gene would be too low by more than an order of magnitude to encode the chemical structure it is known to determine. The insights of molecular genetics reached during the Dogmatic Period readily resolve this apparent paradox: the protein molecule is not assembled by soldering together a kit consisting of 4000 marked atoms of which each atom is potentially connectable to every other atom. Instead the protein assembly proceeds by joining together a specific sequence of 300 amino acid building blocks, each containing a dozen or so atoms each, selected from a pool of 20 different kinds of amino acids. Thus to specify this assembly process only $300 \times \log_2 20$ or 1300 bits of information are needed, or less than the maximum information content of the correspondent gene. This example shows, therefore, that until the process is known by which the nervous system comes into being, or until some detailed credible algorithms have been developed, it is impossible to form even a rough estimate of the amount of genetic information that would be needed to specify the neuronal network. Hence the possibility that the structure of the nervous system is genetically determined cannot be ruled out on purely information-theoretical grounds. In any case, it is obvious that the ontogenetic nervous development is *not* formally equivalent to assembling a kit of 10^{10} individually labeled nerve cells, all potentially interconnectable, by wiring them together according to a schematic supplied by the manufacturer. This is the second error underlying the spurious antigenetic argument.

*   *   *

Nevertheless, on further consideration of the question of how the genes determine behavior, we reach the insight that—for other than information-theoretical reasons—the network cannot be precisely preprogrammed by the genetic information, because of what Waddington has called "developmental noise". On the one hand, the components of the network must be connected with a high degree of precision in order that the whole system can function with the necessary degree of accuracy and reliability. But, on the other hand, there obviously exists an intrinsic rate of error and uncertainty in the epigenetic realization of not only these component cells and their connections but also of other nonnervous body parts with which the nervous system must interact. Thus to realize the required precision, the development of the nervous system can be governed only by a rough program under which there arises an overproduction of cells and connections, from among which an appropriate subset is selected by various testing procedures.

By way of example of this principle, we may consider the development of the visual pathway of the cat whose DNA nucleotide sequence Carl Sagan proposed sending via radio to the Alien Civilization. In the discussion of the preceding chapter regarding the data-abstraction process carried out in that visual pathway, no mention was made of the binocular aspects of vision. But we shall now take into account that the optical system of the cat (as does that of humans) allows both eyes to see the same visual field. In order to provide for fusion of the binocular visual input of the same scene into a single visual percept, each "simple" or "complex" cell of the visual cortex of the cat brain receives electrical signals originating from matched sets of a few thousand primary light receptors in the retinas of both eyes. These sets are matched in the sense that they receive their light from exactly the same points in the visual space.

How do these retino-cortical connections arise during embryonic development of the cat? We might envisage that there is some, as yet unknown, gene-determined process which directs the formation of nerve cell outgrowths and contacts in such a manner that light receptor cells from the corresponding areas of right and left retinas connect to the same cortical nerve cell. But here we must take into account that for binocular vision the "correspondence" of retinal areas depends not only on the topography of the retina but also on the physical optics of the eye. That is to say, which pair of retinal light receptors in the right and left eye happen to see the same point in the visual space is governed by the exact structure and positioning of right and left lenses. Although the epigenetic realization of the physical optics might also occur by gene-determined processes, it is well-nigh inconceivable that the independent formation of the retinas and the lenses could be genetically preprogrammed to occur with such a high degree of precision that the image of a given point in the visual field always falls exactly on that pair of light receptor cells which the genes have managed to connect to the same cortical cell.

As neurological studies of the development of the retino-cortical connections of the cat have revealed, this perplexing developmental problem is solved by providing for a (possibly genetically determined) *overconnection* of light receptor and cortical nerve cells. That is to say, at birth, prior to visual experience, each cortical cell is connected to light receptor cells from a much larger retinal area than is actually compatible with sharp vision. This imprecise congenital visual system is then refined by early postnatal visual experience of the kitten, by a neurophysiological process that identifies those corresponding retinal areas of the two eyes which, given the actual physical optics which the young animal happens to have, do receive light from the same point in the visual space. Thanks to this identification, the developing nervous system selects among the excess of existing retino-cortical connections just those which bring to each binocular cortical cell a coherent visual input.

* * *

Although it thus appears that the genes cannot precisely preprogram the connections of the nervous system, it is nevertheless obvious that they must play some considerable role in the genesis of its structure. And, hence the genes would also make an important contribution to the determination of an animal's behavior. The recognition of this fact has given rise to a neurobiological specialty which aims at discovering just how the genes perform this determinative epigenetic function, and which is a subfield of the more general discipline known as "developmental genetics."

The principal approach used thus far in the effort to establish the role which the genes play in the determination of nerve cell connections is to isolate gene mutations affecting the behavior of an animal and noting changes in the structure of the nervous system responsible for the altered behavior. This approach is evidently based on the belief that the procedure of isolating mutants and analyzing the resulting abnormalities, which molecular geneticists used with such brilliant success in the elucidation of the explicit meaning of the genetic information, will also be of service in unraveling its implicit meaning. This belief is undoubtedly correct, since what Waddington called the "remodeling of the epigenetic landscape" by a gene mutation and its attendant changes can help to identify the functional relations that produce the normal developmental pathways. But it must be borne in mind that though a mutated gene may help to identify a particular epigenetic function, the connection between that function and the mutant gene can be very indirect and involve many other members of the functional network. In view of their general remoteness from the primary action of the genes, the great majority of epigenetic algorithms are unlikely to refer to any gene at all.

In order to appreciate the kind of insights into the development of the nervous system that this genetic approach is likely to provide we may now consider, as a paradigmatic case, the Siamese cat, which happens to carry a mutation that affects its behavior and produces anatomically and physiologically identifiable changes in the nervous system.

The visual pathway of the cat is arranged such that the cerebral cortex on the right side of the animal receives visual input only from the left half of the visual space and the left cerebral cortex only from the right half of the visual space. To produce this right-left crossover of the visual input, the optic nerve fibers reporting from light receptor cells located in the *nasal* half of the retina (i.e., the half which is next to the nose and which receives light from the same side of the visual space as the side of the body on which the eye is located) cross over to the cerebral cortex on the other side of the body, whereas the optic nerve fibers reporting from light receptors located in the *temporal* half of the retina (i.e., the half which is next to the temples and which receives light from the opposite side of the visual space) do not cross over and connect to the cortex on the same side of the body. In normal felines, i.e., in the ordinary domestic cat, the line of demarcation for crossing over of optic nerve fibers is exactly midway between the nasal and temporal edges of the retina. However, as was discovered by R. W. Guillery, in Siamese cats the normal line of demarcation is displaced from the midline toward the temporal edge of the retina. As a result of this displacement, some optic nerve fibers reach the cerebral cortex on the "wrong" side of the brain.

It is beyond the intended scope of this discussion to consider in detail the extremely interesting changes which this faulty cerebral projection of the optic nerve fibers produces in the nervous system and behavior of the Siamese cat. Suffice it to say that in response to the aberrant visual input both the cerebral cortex and the behavior of these animals is reorganized in specific and functionally obviously adaptive ways, so as to minimize the pernicious effect of this genetically determined developmental malformation. Moreover, the nature of this cerebral reorganization lends strong support to our previous conclusion that in the course of development of the nervous system the final connections depend partly on a selective process based on the functional testing of a tentative, imprecise circuitry. What I do want to consider here, however, is the way in which the mutation of a gene carried by the cat helps us discern the genetic component of behavior.

*   *   *

Although the informational content of the genome of the Siamese cat certainly differs from that of the ordinary cat in more than one gene, Guillery has been able to identify the mutant gene whose change is responsible for the aberrant crossover of the optic nerve fibers. It is the "tyrosinase" gene in which the amino acid sequence of a protein is inscribed that catalyzes a reaction step in the biosynthesis of the dark pigment melanin. In the Siamese cat this gene carries a mutation which renders the mutant protein unable to carry out its catalytic function at 37°C and thus prevents the synthesis of the dark pigment at body temperature. It is this mutation which is responsible for the characteristic Siamese coat color, namely the lightly pigmented body fur framed by black hair patches on the tips of the ears, the paws and the snout.

But what is the possible connection between the formation of melanin and the directed outgrowth of the optic nerve fibers from the retina to the right or left cerebral cortex? And why does the absence of pigment produce an aberrant crossing over, particularly in view of the fact that the retinal cells that provide the optic nerve fibers do not normally contain significant amounts of melanin in any case? The answers to these questions are not yet available, but it is not difficult to invent a set of hypothetical epigenetic algorithms that would provide a plausible formal explanation. For this purpose, we envisage that the embryonic retinal nerve cells have some property whose measure x increases monotonically across the retina from the temporal to the nasal edge. We envisage, furthermore, that all those cells for which $x > x_0$ send their optic nerve fibers to the opposite side of the brain, with the remainder of the cells sending their fibers to the same side. The normal developmental system (i.e., the retinal region of the epigenetic landscape) is so poised that $x = x_0$ at the retinal midline. Now although the retinal nerve cells do not themselves contain melanin, they are, in fact, the direct developmental descendants of the layer of melanin-containing cells that form the pigmented epithelium which lies behind the retina and shields it from stray light. To complete our sample algorithms we need merely envisage that if the epithelial precursor cells are not pigmented, as they are not in the Siamese cat embryo, the normal developmental schedule of formation of their retinal nerve cell descendants is slightly perturbed. And, as a consequence of this slight remodeling of the epigenetic landscape, the temporal-nasal retinal gradient of the postulated property is also slightly perturbed in such a manner that x reaches the value x_0 already on the temporal side of the retinal midline. Thus the absence of pigment in the epithelial precursor cells would cause those optic nerve fibers reporting from light receptors in the temporal half of the retina for which $x>x_0$ to make an "incorrect" outgrowth to the cerebral cortex on the opposite side of the brain.

* * *

We are now in a position to understand the sense in which the "tyrosinase" gene in the feline genome takes part in the "determination" of the visual pathway and behavior: it directs the assembly of a particular protein molecule from its constituent amino acids in the precursor cells of the optic nerve fibers. The presence of the pigment whose synthesis is catalyzed by that protein is a necessary condition for the "normal" development of the pathway, in that the absence of the pigment from the precursor cells sets off a cascade of dysfunctional, albeit specific aberrations, which eventually leads to a profound reorganization of a part of the brain of the animal. The visual pathway of the Siamese cat is, therefore, an ideal case for the genetic approach: a mutation in a single known gene that determines the amino acid sequence of a known protein that catalyzes a known chemical reaction whose end product has a known physiological function causes a known specific and striking structural change of the nervous system. But, alas, it tells us rather little about the *explicit* genetic component of behavior that we did not know already. Rather the tremendous interest of the Siamese cat lies in the help it is likely to provide in the effort to discover the epigenetic algorithms that govern the contextual realization of the *implicit* meaning of the genetic information. For instance, the probable connection between the absence of retinal epithelial pigment and the misdirection of the optic nerve fibers suggests some testable hypotheses about the rules that determine whether an optic nerve fiber grows to the same or to the opposite side of the brain as its retina of origin. The eventual statement of these hypothetical rules may contain such terms as enzymes, gradients, growth rates, threshold concentrations, preferential adhesion and nerve impulse frequencies, but the word "gene" is unlikely to find frequent mention.

It is possible, of course, that the genetic approach may yet lead to the discovery of DNA nucleotide sequences whose explicit meaning is, in fact, *directly* concerned with the determination of the structure of the nervous system. And if such genes do turn up, the information they could provide would undoubtedly be of tremendous help in our effort to discover the developmental algorithms. But, all the same, it seems most likely that the great majority of mutants isolated for abnormal behavior will manifest changes in their nervous system because the mutation has occurred in a gene whose meaning bears implicitly rather than explicitly on the functional relations that form the epigenetic landscape.

In the light of these considerations we can try to appreciate the nature of the contribution which the genetic approach can make to the understanding of the nervous system. There can be no doubt that the genetic approach is of great practical and technical significance. First, within the context of human psychology and medicine it is of the utmost importance to understand the hereditary component of normal or pathological behavior. For instance, if it could be shown that schizophrenia is "determined" by a particular mutant gene, then the value of this knowledge would be in no way diminished by the realization that a vast contextual hierarchy separates the explicit molecular biological meaning of that gene from its implicit epigenetic meaning for behavior. Second, within the context of neurophysiology the method of "genetic dissection" of behavior is likely to be of great assistance for functional analysis of known nerve cell networks. For instance, an abnormal behavior and a concomitant abnormal structure of the nervous system of a mutant genotype can obviously provide insights into how the normal circuitry generates the normal behavior. Third, within the context of developmental biology the remodeling of the epigenetic landscape by mutant genes can, as we saw in the case of the Siamese cat, help us to recognize the functional relations that create the normal pathways that lead to the endpoint of the adult animal. But as far as the discovery of how the genes interact with the environment to determine behavior is concerned, we can see that, thanks to the past achievements of molecular biology, that discovery has already been made: the genes determine behavior, just as they determine any other aspect of the phenome, by directing the synthesis of specific proteins.

Thus the deep biological problem in want of a solution is not how genes determine behavior but to find the algorithms of the living world that produce the nervous system. The horizon of that discipline, which might be called "Neurological Epigenetics," lies far beyond the genes, and encompasses the context under which the explicit meaning of the genetic information gives rise to the implicit meaning that is the organism.

Bibliography

Benzer, Seymour. "From Gene to Behavior." *J. American Medical Association* 218, 1015–1022 (1971).

Benzer, Seymour. "The Genetic Dissection of Behavior." *Scientific American*, December 1973, pp. 24–37.

Brenner, Sydney. "The Genetics of Behavior." *British Medical Bulletin* 29(3), 269–271 (1973).

Guillery, R. W. "Visual Pathways in Albinos." *Scientific American*, May 1974, p. 44–54.

Guillery, R. W., V. A. Casagrande, and M. D. Oberdorfer. "Congenitally Abnormal Vision in Siamese Cats." *Nature* 252, 195–199 (1974).

Gurdon, J. B. "Transplanted Nuclei and Cell Differentiation." *Scientific American*, December 1968, pp. 24–35.

Horridge, G. A. *Interneurons.* San Francisco: W. H. Freeman and Company, 1968.

Jacob, François. *The Logic of Life.* New York: Pantheon, 1973.

Markert, C. L., and H. Ursprung. *Developmental Genetics.* Englewood Cliffs, N.J.: Prentice Hall, 1971.

Sagan, Carl. *Communication With Extraterrestrial Intelligence.* Cambridge, Mass.: M.I.T. Press, 1973.

Schrödinger, E. *What Is Life?* New York: Cambridge University Press, 1945.

Waddington, C. H. *The Strategy of the Genes.* London: George Allen & Unwin, 1957.

Weismann, August. *The Germ Plasm*. Translated by W. N. Parker and Harriet Ronnfeldt. London: Walter Scott, 1893.

Wolff, Caspar F. *Theorie der Generation*, 1759. Ostwald's Klassiker der Exakten Wissenschaften, No. 84, 1896.

Woodger, J. H. "What do we mean by inborn?" *British Journal of the Philosophy of Science* 3, 319 (1953).

LXXVIII.
Comment
vne idée
peut eftre
cõpofée de
plufieurs;&
d'où vient
qu'alors il
ne paroist
qu'vn feul
objet.
Et de plus, pour entendre icy par occafion, comment, lors que les deux yeux de cette machine, & les organes de plufieurs autres de fes fens font tournez vers vn mefme objet, il ne s'en forme pas pour cela plufieurs idées dans fon cerveau, mais vne feule, il faut penfer que c'eft toujours des mefmes points de cette fuperficie de la glande H que fortent les Efprits, qui tendant vers divers tuyaux peuvent tourner divers membres vers les mefmes objets: Comme icy que c'eft du feul point b que fortent les Efprits, qui tendant vers les tuyaux 4, 4, & 8, tournent en mefme temps les deux yeux & le bras droit vers l'objet B.

Descartes' theory of visual perception, as published posthumously in his Traité de l'Homme *(1667). The pear-shaped cerebral structure labeled "H" is the pineal gland, thought by Descartes to be the gateway to the soul, where the percept is formed. Thus, according to this view, it is to the pineal gland that information from the input part projects and it is from the pineal gland that commands to the effector part issue. [From the Kofoid collection of the Biology Library of the University of California, Berkeley. Courtesy of the University of California.]*

10

LIMITS
TO THE
SCIENTIFIC
UNDERSTANDING
OF MAN

[*1975*]

For the past two centuries scientists, particularly in English-speaking countries, have generally viewed their attempt to understand the world from the epistemological viewpoint of positivism. All the while, positivism had been under attack from philosophers, but it is only since the 1950s that its powerful hold on the students of nature finally seems to be on the wane. There is as yet no generally accepted designation for the philosophical alternatives that are replacing positivism, but the view of man known as "structuralism" appears to be central to the latter-day epistemological scene.* But whereas the work of structuralist scientists has shown up the essential barrenness of the positivist approach to human behavior, even the structuralist program, however meritorious, is unlikely to lead to a scientifically validated understanding of man.

---

*An excellent overview of the structuralist movement can be obtained from H. Gardner, *The Quest for Mind* (Knopf, New York, 1973).

Positivism

The principal tenet of positivism, as formulated in the eighteenth century mainly by David Hume and the French Encyclopedists, is that, since experience is the sole source of knowledge, the methods of empirical science are the only means by which the world can be understood.* According to this view, the mind at birth is a clean slate on which there is gradually sketched a representation of reality built on cumulative experience. This representation is orderly, or structured, because, thanks to the principle of inductive reasoning, we can recognize regular features of our experience and infer causal connections between events that habitually occur together. The possibility of innate or a priori knowledge of the world, a central feature of the seventeenth-century rationalism of René Descartes, is rejected as a logical absurdity.

It is unlikely that the widespread acceptance of positivism had a significant effect on the development of the physical sciences, since physicists have little need to look to philosophers for justification of their research objectives or working methods. Moreover, once a physicist has managed to find an explanation for some phenomenon, he can be reasonably confident of the empirical test of its validity. For instance, the positivist rejection of the atomic theory in the late nineteenth century, on the grounds that no one had ever "seen" an atom, did not stop chemists and physicsts from then laying the groundwork for our present understanding of microscopic matter. However, in the human sciences, particularly in psychology and sociology, the situation was quite different. Here positivism was to have a most profound effect. One reason for this is that practitioners of the human sciences are much more dependent on philosophical support of their work than are physical scientists. In contrast to the clearly definable research aims of physical science, it is often impossible to state explicitly just what it really *is* about human behavior that one wants to explain. This in turn makes it quite difficult to set forth clearly the conditions under which any

---

*I am referring to Hume as a founder of positivism, even though the name of that philosophical view was invented much later by Auguste Comte, because he shaped seventeenth-century empiricism into the anti-metaphysical outlook that informed much of the nineteenth- and twentieth-century science.

postulated causal nexus linking the observed facts could be verified. Nevertheless, positivism helped to bring the human sciences into being in the first place, by insisting that any eventual understanding of man must be based on the observation of facts, rather than on armchair speculations. But, by limiting inquiry to such factual observations and allowing only propositions that are based on direct inductive inferences from the raw sensory data, positivism constrained the human sciences to remain taxonomic disciplines whose contents are largely descriptive with little genuine explanatory power. Positivism clearly informed the nineteenth-century founders of psychology, ethnology, and linguistics. Though we are indebted to these founders for the first corpus of reliable data concerning human behavior, their refusal to consider these data in terms of any propositions not derived inductively from direct observation prevented them from erecting a theoretical framework for understanding man.

Structuralism

Structuralism transcends the limitation on the methodology, indeed on the agenda of permissible inquiry, of the human sciences imposed by positivism. Structuralism admits, as positivism does not, the possibility of innate knowledge not derived from direct experience. It represents, therefore, a return to Cartesian rationalist philosophy. Or, more exactly, structuralism embraces this feature of rationalism as it was later reworked by Immanual Kant for his philosophy of critical idealism. Kant held that the mind constructs reality from experience by use of innate concepts, and thus to understand man it is indispensable to try to fathom the nature of his deep and universal cognitive endowment. Accordingly, structuralism not only permits propositions about behavior that are not directly inducible from observed behavioral data, but it even maintains that the relations between such data, or *surface structures*, are not by themselves explainable. According to this view the causal connections that determine behavior do not relate to surface structures at all. Instead, the overt behavioral phenomena are generated by covert *deep structures*, inaccessible to direct observation. Hence any theoretical framework for understanding man must be based on the deep structures, whose discovery ought to be the real goal of the human sciences.

Probably the best known pioneer of structuralism is Sigmund Freud, to whom we owe the fundamental insight that human behavior is governed not so much by the events of which we are consciously aware in our own minds or which we can observe in the behavior of others, but rather by the deep structures of the subconscious which are generally hidden from both subjective and objective view. The nature of these covert deep structures can only be inferred indirectly by analysis of the overt surface structures. This analysis has to proceed according to an elaborate scheme of psychodynamic concepts that purports to have fathomed the rules which govern the reciprocal transformations of surface into deep and of deep into surface structures. The great strength of Freudian analytical psychology is that it does offer a theoretical approach to understanding human behavior. Its great weakness, however, is that it is not possible to verify its propositions. And this can be said also of most other structuralist schools active in the human sciences. They do try to explain human behavior within a general theoretical framework, in contrast to their positivist counterparts who cannot, or rather refuse to try to do so. But there is no way of verifying the structuralist theories in the manner in which the theories of physics can be verified through critical experiments or observations. The structuralist theories are, and may forever remain, merely plausible, being, maybe, the best we can do to account for the complex phenomenon of man.

Ethnology and Linguistics

For instance, positivist ethnology, as conceived by one of its founders, Franz Boas, sought to establish as objectively and as free from cultural bias as possible the facts of personal behavior and social relations to be found in diverse ethnic groups. Insofar as any explanations are advanced at all to account for these observations, they are formulated in functionalist terms. That is to say, every overt feature of behavior or social relation is thought to serve some useful function in the society in which it is found. The explanatory work of the ethnologist would be done once he has identified that function and verified its involvement by means of additional observations. Accordingly, the general aim of this approach to ethnology is to show how manifold and diverse the ways are in which man has adapted his behavior and social existence to the range of conditions that he encountered in settling the earth. By contrast, structuralist ethnology, according to one of its main exponents, Claude Lévi-Strauss, views the concept of functionality as a tautology, devoid of any real explanatory power for human behavior. All extant behavior is obviously "functional" since all "disfunctional" behavior would lead to the extinction of the ethnic group that exhibits it. Instead of functionality, so Lévi-Strauss holds, only universal and permanent, deep structural aspects of the mind can provide any genuine understanding of social relations. The actual circumstances in which different peoples find themselves no more than modulate the overt behavior to which the covert deep structures give rise. In other words, the point of departure of structuralist ethnology is the view that the apparent diversity of ethnic groups pertains only to the surface structures and that at their deep structural level all societies are very much alike. Hence the general aim of structuralist ethnology is to discover those universal, deep mental structures which underlie all human customs and institutions.

Positivist linguistics, as conceived by its founders, such as Ferdinand de Saussure and Leonard Bloomfield, addresses itself to the discovery of structural relations among the elements of spoken language. That is to say, the work of that school is concerned with the surface structures of linguistic performance, the patterns which can be observed as being in use by speakers of various languages. Since the patterns which such classificatory analysis reveals differ widely, it seems reasonable to conclude that these patterns are arbitrary, or purely conventional, one linguistic group having chosen to adopt one, and another group having chosen to adopt another convention. There would be nothing that linguistics could be called on to explain, except for the taxonomic principles that account for the degree of historical relatedness of different peoples. And if the variety of basic patterns of various human languages is indeed the result of arbitrary conventions, study of extant linguistic patterns is not likely to provide any deep insights into any universal properties of the mind. By contrast, the structuralist approach to linguistics, according to its main modern proponent, Noam Chomsky, starts from the premise that linguistic patterns are not arbitrary. Instead, all men are believed to possess an innate, a priori knowledge of a universal grammar, and that despite their superficial differences, all natural languages are based on that same grammar.* According to that view, the overt surface structure of speech, or the organization of sentences, is generated by the speaker from a covert deep structure. In his speech act, the speaker is thought to generate first his proposition as an abstract deep structure that he transforms only secondarily according to a set of rules into the concrete surface structure of his utterance. The listener in turn fathoms the meaning of the speech act by just the inverse transformation of surface to deep structure. Chomsky holds that the grammar of a language is a system of transformational rules that determines a certain pairing of sound and meaning. It consists of a syntactic component, a semantic component, and a phonological component. The surface structure contains the information relevant to the phonological component, whereas the deep structure contains the information relevant to the semantic component, and the syntactic component pairs surface and deep structures.

So far, it does not seem to have been possible to identify clearly those aspects of the grammar of any one natural language which are universal, and hence shared with all other natural languages, in contrast to those aspects which are peculiar, and hence responsible for differentiating that language from other languages. Some success has been achieved at the sound level, where a limited number of universal "distinctive features" has been identified. Each feature takes on one of a very few discrete values (e.g. "present" or "absent") in a given

---

*Unfortunately, calling Chomsky a "structuralist" is bound to raise confusion. Among students of linguistics, Chomsky is called a "generative grammarian," whereas not he but his positivistic predecessors, whose limited goals Chomsky has sought to transcend, are known as "structuralists." But in view of Chomsky's evident philosophical affinity to Freud, Lévi-Strauss, and other "structuralist" contributors to the human sciences, there seems to be no way to avoid this terminological confusion in a general discussion of his position.

sound element of speech. In other words, every symbol of a phonetic alphabet can be regarded as a set of these features, each with a specified value. Thus it should be possible to construct a universal phonetic script which would allow, in principle at least, a speaker of any natural language to pronounce correctly a written text in any other natural language. Much less success has been achieved so far at the philosophically more interesting meaning level. Here the concept of a universal grammar would suggest the existence of an ensemble of universal semantic "distinctive features" and laws regarding their interrelations and permitted variety. That is to say, every meaningful concept would be fathomable as a set of semantic features, each with a specified value. From this point of view, it should be possible to construct a universal "semantic script," texts of which all speakers of natural languages would understand. Unfortunately, it has proven difficult to put forward any specific proposals for or examples of the putative "semantic features," except to conclude that they must be of a highly abstract nature. In any case, if both the surface level of sound and the deep level of meaning are universal aspects on which all natural languages are based, then it must be the transformational components of grammar that have become greatly differentiated during the course of human history, since the building of the Tower of Babel. But the presumed constancy through time of the universal aspects cannot be attributable to any cause other than innate, hereditary aspects of the mind. Hence, the general aim of structuralist linguistics is to discover those universal aspects.

Transcendental Concepts

Now, in retrospect, at a time when positivism and its philosophic and scientific ramifications appear to be moribund, it seems surprising that these views ever did manage to gain such a hold over the human sciences. Hume, though one of its founders, already saw that the positivist theory of knowledge has a near-fatal logical flaw. As he noted, the validity of inductive reasoning—which, according to positivism forms the basis of our knowledge of the regularity of the world, and hence for our inference of causal connections between events—can neither be demonstrated logically nor can it be based on experience. Instead, inductive reasoning is evidently something that man brings to rather than derives from experience. Not long after Hume, Kant showed that the positivist doctrine that experience is the sole source of knowledge derives from an inadequate understanding of the mind. Kant pointed out that sensory impressions become experience, that is, gain meaning, only after they are interpreted in terms of a priori concepts, such as time and space. Other a priori concepts, such as induction (or causality), allow the mind to construct reality from that experience. Kant referred to these concepts as "transcendental," because they transcend experience and are thus beyond the scope of scientific inquiry. But why was it that, although Kant wielded an enormous influence among philosophers, his views had little currency among scientists? Why did positivism rather than Kant's "critical idealism" come to inform the explicit or implicit epistemological outlook of much of nineteenth- and twentieth-century science? At least two reasons can be advanced for this historical fact. The first reason is simply that many positivist philosophers, especially Hume, were lucid and effective writers whose message could be readily grasped after a single reading of their works. The texts of Kant, and of his mainly Continental followers, are, by contrast, obscure and hard to understand.

The second reason for the long scientific neglect of Kant is more profound. After all, it does seem very strange that if, as Kant alleges, we bring such concepts as time, space, and causality to sensation a priori, these transcendental concepts happen to fit our world so well. Considering all the ill-conceived notions we might have had about the world prior to experience, it seems nothing short of miraculous that our innate concepts just happen to be those that fill the bill.* Here the positivist view that all knowledge is derived from experience a posteriori seems much more reasonable. It turns out, however, that the way to resolve the dilemma posed by the Kantian a priori has been open since Charles Darwin put forward the theory of natural selection in the mid-nineteenth century. Nevertheless, few scientists seem to have noticed this until Konrad Lorenz drew attention to it 30

---

*Kant himself rejected the only resolution of this dilemma available at his time, namely, that it was God who put these concepts into man's mind.

years ago. Lorenz pointed out that the positivist argument that knowledge about the world can enter our mind only through experience is valid if we consider only the ontogenetic development of man, from fertilized egg to adult. But once we take into account also the phylogenetic development of the human brain through evolutionary history, it becomes clear that individuals can also know something of the world innately, prior to and independent of their own experience. After all, there is no biological reason why such knowledge cannot be passed on from generation to generation via the ensemble of genes that determines the structure and function of our nervous system. For that genetic ensemble came into being through the process of natural selection operating on our remote ancestors. According to Lorenz, "experience has as little to do with the matching of a priori ideas with reality as does the matching of the fin structure of a fish with the properties of water." In other words, the Kantian notion of a priori knowledge is not implausible at all, but fully consonant with present mainstream evolutionary thought. The a priori concepts of time, space, and causality happen to suit the world because the hereditary determinants of our highest mental functions were selected for their evolutionary fitness, just as were the genes that give rise to innate behavioral acts, such as sucking the nipple of mother's breast, which require no learning by experience.

The importance of these Darwinian considerations transcends a mere biological underpinning of the Kantian epistemology. For the evolutionary origin of the brain explains not only why our innate concepts match the world but also why these concepts no longer work so well when we attempt to fathom the world in its deepest scientific aspects.

As was set forth in earlier chapters, this barrier to unlimited scientific progress posed by the a priori concepts which we necessarily bring to experience was a major philosophical concern of Bohr. Bohr had recognized that the basis of scientific thought and communication is our ordinary, everyday language and that the tremendous increase in the range of our experience has put in question the sufficiency of concepts and ideas incorporated in that language. The most basic of these concepts and ideas are precisely the Kantian a priori notions of time, space, and causality. The meaning of these terms is intuitively obvious and grasped automatically by every child in the course of its normal intellectual development, without the need to attend physics classes. Accordingly, the models that modern science offers as explanations of reality are pictorial representations built of these intuitive concepts. We can now see why this procedure was satisfactory as long as explanations were sought for phenomena that are commensurate with the events that are the subject of our everyday experience. For it was precisely for its fitness to deal with everyday experience that our brain was selected in the evolutionary sequence that culminated in the appearance of *Homo sapiens*. But when, at the turn of this century, physicists began to study tiny subatomic or immense cosmic events, serious conceptual difficulties arose because our mental equipment was not selected for dealing successfully with phenomena so far removed dimensionally from the experiential realm. Thus a Darwinian explanation can be readily advanced for Bohr's epistemological discovery that the enormous enlargement of the scope of science brought about by twentieth century physics was achieved only at the price of denaturing the intuitive meaning of some of its basic concepts with which man starts out in his quest for understanding nature.

The Grandmother Cell

In addition to explaining in evolutionary terms how the human brain and thus its isomorph, the mind, can gain possession of a priori concepts that match the world, modern biology has shown also that the brain does appear to operate according to principles that correspond to the tenets of structuralism. This is not to imply that the neurological correlates of any of the structuralist notions, particularly not of Freud's subconscious, or of Lévi-Strauss' ethnological universals, or of Chomsky's universal grammar, have actually been found. Such a claim would be nonsensical, inasmuch as it is not even known in which parts of the brain the corresponding processes occur. But as was set forth earlier in Chapter 8, Hubel and Wiesel's neurobiological studies of the visual pathway have indicated that, in accord with the structuralist tenets, information about the world reaches the depths of the mind, not as raw data but as highly processed structures that are generated by a set of stepwise, preconscious informational transformations of the sensory input. These cerebral transformations proceed according to a program that pre-exists in the brain. These findings thus lend biological support to the structuralist dogma that explanations of behavior must be formulated in terms of such deep programs and reveal the wrong-headedness of the positivist approach, which rejects the postulation of covert internal programs as "mentalism."

At this point, however, we must return to a question raised but not answered in Chapter 8, where we first considered the discovery that the visual pathway subjects the sensory input to the primary light receptors to a stepwise abstraction process in which information is selectively destroyed. We then asked how far this process of cellular abstraction by convergence of neural pathways on more and more "meaningful" individual neurons could be imagined to go. Should we suppose that the cellular abstraction process goes so far that there exists for every meaningful structure of whose specific recognition a person is capable (for example, "my grandmother") at least one particular nerve cell in the brain that responds when and only when the light and dark pattern from which that structure is abstracted appears in its visual space.*

This could very well be the case for lower animals, with their limited behavioral repertoire. For instance, neurobiological evidence secured by J. Y. Lettvin, H. R. Maturana, and W. H. Pitts has shown that the visual system of the frog abstracts its input data in such a way as to produce only two meaningful structures, "my prey" and "my predator," which, in turn, evoke either of two alternative behavioral outputs, attack or flight. But in the case of man, with his vast semantic capacities, this picture does not appear very plausible, despite the fact that the human brain has many more nerve cells than the frog's brain. Somehow, for man the notion of the single cerebral nerve cell as the ultimate element of meaning seems worse than a gross oversimplification; it seems qualitatively wrong. Yet, so far at least, it

---

*A fuller discussion of the crucial importance of this question for an eventual neurological account of perception is given by H. B. Barlow, *Perception* 1, 371 (1972).

is the only neurologically coherent scheme that can be put forward. Admittedly, ever since the discipline of neurobiology came into being more than a century ago, there have been adherents of a "holistic" theory of the brain. This theory envisions that specific functions of the brain, including perception, depend not on the activity of particular localized cells or centers but on general and widely distributed activity patterns. Such holistic theories, however, amount to little more than phenomenological recapitulations of neurological correlates of behavior or mental activity. They are not, therefore, explanatory in a scientific sense. This does not mean that the holistic approach to the brain is necessarily wrong; it merely means that it concedes at the outset that the brain cannot be explained.

The Self

We thus encounter the barrier to an ultimate scientific understanding of man which Descartes had recognized more than three centuries ago. Descartes had clearly outlined the nature of the problem posed by vision, and the modern neurological findings mentioned in the preceding paragraphs represent latter-day triumphs of the Cartesian approach. At the same time, Descartes had realized that physiological studies really leave the central problem of visual perception untouched. For the percept is obviously a function of the *soul*, or in modern psychological parlance, of the *self*, whose nature Descartes thought to be inaccessible to scientific analysis. No matter how deeply we probe into the visual pathway, in the end we need to posit an "inner man" who transforms the visual image into a percept. And, as far as linguistics is concerned, the analysis of language appears to be heading for the same conceptual impasse as does the analysis of vision. I think it is significant that Chomsky, who views himself as carrying on the line of linguistic analysis begun by Descartes and his disciples, has encountered difficulty with the postulated semantic component. Thus far, it has not been possible to spell out how the semantic component manages to extract meaning from the informational content of the deep structure. It is over just the problem of meaning that disputes have arisen between Chomsky and some of his students, and it does not seem that any solution is at present in sight. As pointed out by John Searle in his appreciation of "Chomsky's Revolution in Linguistics," the obstacle in the way of giving a satisfactory account of the semantic component appears to reside in defining explicitly the problem that is to be solved. That is to say, for man the concept of "meaning" can be fathomed only in relation to the self, which is both ultimate source and ultimate destination of semantic signals. But the concept of the self, the cornerstone of Freud's analytical psychology, cannot be given an explicit definition. Instead, the meaning of "self" is intuitively obvious. It is another Kantian transcendental concept, one which we bring a priori to man, just as we bring the concepts of space, time, and causality to nature. The concept of self can serve the student of man as long as he does not probe too deeply. However, when it comes to explaining the innermost workings of the mind—the deep structure of structuralism—then this attempt to increase the range of understanding raises, in Bohr's terms, "questions as to the sufficiency of concepts and ideas incorporated in daily language." Thus, the image of man as a Russian doll, with the outer body encasing an incorporeal inner man, is evidently a presupposition hidden in the rational linguistic use of the term "self," and the attempt to eliminate the inner man from the picture only denatures that intuitive concept beyond the point of psychological utility.* From this ultimate insufficiency of the everyday concepts which our brain obliges us to use for science it does not, of course, follow that further study of the mind should cease, no more than it follows from it that one should stop further study of physics. But I think that it is important to give due recognition to this fundamental epistemological limitation to the human sciences, if only as a safeguard against the psychological or sociological prescriptions put forward by

---

*Thus I reject S. Toulmin's claim (in *The Neurosciences*, G. C. Quarton, T. Melnechuck, F. O. Schmidt, eds. Rockefeller Univ. Press, New York, 1967, p. 822) that the picture of the inner man is merely a legacy of the applications of seventeenth-century physics to the study of man and that the need for the concept vanishes within the frame of reference of twentieth-century physics. Referring to an entirely different nonscientific tradition, we may note that the *satori* of Zen Buddhism demands purging the mind of its innate self concept and that the deep insights into man gained thereby cannot consequently be communicated by explicit verbal discourse.

those who allege that they have already managed to gain a scientifically validated understanding of man.*

* * *

Postscript (1978). Among the letters to the editor of *Science* commenting on this essay after its original appearance were two that criticized me for not mentioning the contributions of Jean Piaget, one of the leading figures of the structuralist movement. I replied that even though I could refer to only a tiny fraction of the large number of contributors to the vast subject of the scientific understanding of man, I agreed, in retrospect, that I *ought* to have mentioned Piaget's important studies of cognitive development in children. Kant's claim that our fundamental concepts of time and space are a priori, and hence immanent in human reason, does not necessarily mean that they are already present, full blown, at birth. On the contrary, according to Piaget, they are *not* present at birth and are built up only gradually during childhood as a result of an orderly process of "genetic epistemology." This process of gradual construction of the elements of rational thought passes through a series of clearly recognizable stages and depends on sensorimotor interactions of the child with its environment. For instance, Piaget found that at an early stage the infant first builds elementary forms of concrete classificatory and relational systems, such as the notion of an object with a characteristic state, which at a yet earlier stage it still lacks. Only after the child has begun to develop such concrete notions as constant size and identity of the objects of its surround, can it develop more abstract linguistic, logical, and mathematical modes of thought. As far as the Kantian categories of space and time are concerned, Piaget found that they take on their mature form at a relatively late stage, prior to which space

---

*After this article had originally gone to press I came upon *Seelenglaube und Psychologie* (Deuticke, Leipzig, 1930) by Freud's disciple and critic Otto Rank. Using many of the same arguments that I put forward here, including the epistemological parallels between twentieth-century physics and psychology, Rank concluded that the unavoidable concept of the soul limits the eventual scientific understanding of man. I am indebted to A. Wheelis for calling my attention to Rank's book.

and time still appear to be conceptually intertwined. The importance of Piaget's work for this discussion thus lies in his empirical demonstration that our epistemological concepts arise autonomously during infancy and early childhood, as a result of an interaction between the developing nervous system and the world. Hence, just as Lorenz made the Kantian a priori part of modern evolutionary biology, so did Piaget bring it into accord with modern developmental biology and the epigenetic dogma, treated in the preceding chapter, that the phenome arises as a dialectic between the genome and the environment. (See J. Piaget and B. Inhelder, *The Psychology of the Child*, Basic Books, New York, 1969.)

Bibliography

Bohr, Niels. *Atomic Physics and Human Knowledge*. New York: Science Editions, 1961.

Chomsky, N. *Language and Mind*. New York: Harcourt, Brace & World, 1968.

Lettvin, J. Y., H. R. Maturana, W. S. McCulloch, and W. H. Pitts. "What the Frog's Eye Tells the Frog's Brain." *Proc. Inst. Radio Eng.* 47, 1940–1951 (1962).

Lorenz, Konrad. "Kant's Doctrine of the *a priori* in the Light of Contemporary Biology." *in* L. Bertalanffy and A. Rappaport (eds.), *General Systems*. Ann Arbor: Soc. Gen. Systems Research, 1962.

Searle, J. "Chomsky's Revolution in Linguistics." *New York Rev.* 29 June 1972, pp. 16–24.

Immanuel Kant. Portrait by Becker, 1768.

11

THE
DECADENCE
OF
SCIENTISM

[1977]

ɪɴ an article entitled "The Science-Textbook Controversies,"
Dorothy Nelkin pointed out that the latter-day opposition to
elementary and secondary school textbooks that present a Darwinian rather than Biblical approach to biology, and a cultural-
anthropological rather than patriotic-civic approach to social
studies, reflects a current belief in the "decadence of scientism."
According to Nelkin, these textbook critics ought not to be dis-
missed as a mere antiscience fringe group, inasmuch as "they
are not reacting against science so much as resisting its image as
an infallible source of truth that denies their sense of place in the
universe". And just this image is certainly projected by many
scientists. For instance, C. H. Waddington and Julian Huxley
still held in the 1960s that evolution provides a secure basis for
ethics and for a "naturalistic" religion. Indeed, the textbook critics are part of a more general romantic and political resistance
to science, which, as Nelkin warns, may turn out to have painful
consequences. Her article does not mention, however, that the
perception of the decadence of scientism is shared by many
present-day philosophers, who have otherwise little in common
with the Fundamentalist-Christian and Populist textbook crit-

ics. As many contemporary philosophers have pointed out, the claim of scientism that the positive methods and insights of science are valid for the entire sphere of human activity not only lacks philosophical merit but is also politically dangerous: it provides a rational justification for the totalitarian state.

In this final chapter I review some of the shortcomings of scientism, especially in regard to its claim of being able to validate moral action. I hope to show, furthermore, that whereas science cannot itself provide a foundation of ethics, it may be of some use in accounting for moral behavior.

The Rise of Scientism

Until the eighteenth century the foundation of Western ethics rested securely on the divine authority of the Judeo-Christian religion. Then an erosion of religious faith was set in motion by the French Encyclopedists of the Enlightenment, who preached that man's reason rather than God's word provides the authority for moral values. But by the end of the eighteenth century, Immanuel Kant had demonstrated the intrinsic inconsistency of that position: critical reflection on the nature of morality shows that the beliefs in God, freedom, and immortality of the soul are necessary ingredients of any rational system of ethics. As Kant pointed out, if there were no God and no immortality, there would be no counter-argument against the claim that it is irrelevant how we conduct ourselves. Unless there were a God and a life-after-death, there would exist no Supreme Wisdom by whose standards the moral worth of our actions could be judged and no way by which we would eventually know whether we acted well or badly during our temporal sojourn in the here-and-now. Without these beliefs, life could not be other than amoral.

Although Kant's writings wielded an enormous influence on nineteenth century philosophy, they did not ward off the ever-growing dominance of atheistic beliefs. Finally, toward the end of the nineteenth century, Friedrich Nietzsche saw the full implications of that development. He noted that whereas without God mankind cannot lead a moral life, the Encyclopedists and their followers had nevertheless managed to kill God. Thus everything has become permitted, and mankind is about to sink into an amoral abyss. To survive, Nietzsche thought, man must now transcend his animal nature and become truly human, that is to say, to become Superman. He must become his own God, so that his actions will be "beyond good and evil." Though Nietzsche's Superman has yet to make his appearance, we did manage to survive the last 100 years, albeit just by the skin of our teeth. In any case, with Nietzsche's Counter-Enlightenment the eighteenth-century philosophical project to replace the religious by a materialistic basis of ethics reached the end of its road: it had been made clear at last that what the Encyclopedists had in mind for their citizen-atheist was not man but Superman.

In the twentieth century, the philosophical vacuum created by the demise of both traditional Western religiosity and Enlightenment materialism came to be filled by a variety of partly complementary, partly competing approaches to ethics, such as existentialism, psychoanalysis and the ancient Eastern philosophies of Buddhism and Taoism. Nevertheless, there still remain important sectors of the Western world where the message of the Counter-Enlightenment has not yet produced its full impact and where a rear-guard action to salvage the project founded by the Encyclopedists continues in our time. One of these sectors is represented by the Communist countries, where the dialectical materialism of the Enlightenment's most influential disciple, Karl Marx, continues to be maintained as a kind of state religion. Another important sector is represented by the community of contemporary scientists, many of whom are votaries of scientism as an Ersatz religion which professes that the methods of science provide the only kind of authentic knowledge. Since scientism views the traditional theological ground of ethics as a morass of irrational superstitions dating back to a remote, pre-scientific age, it proposes to base an authoritative foundation of ethics on the authority of modern science. However much this scientistic creed may now seem in decadence outside the scientific community, it is still alive and well within it. As we saw in Chapter 6, Crick's *Of Molecules and Men* and Monod's *Chance and Necessity* both represent pleas on its behalf, and scientistic tenets remain the unstated ethical premise conceded implicitly by the opposing sides in debates held currently in scientific circles on matters touching on moral issues.

For an appreciation of the scientistic approach to ethics, it is useful to distinguish two different grades of scientism, namely *hard-core* and *soft-core*. Believers in hard-core scientism take the view that moral norms and values can, or must, be justified on scientific grounds. Believers in soft-core scientism allow that valid moral values can be justified on nonscientific grounds, but they still insist on the primacy of science as a guide to moral action.

Hard-core Scientism

From the objective, scientific point of view *Homo sapiens* is just one of many species in the class of mammals of the phylum of vertebrates of the animal kingdom. Biology, therefore, seems to be the branch of science that is best pressed into service for the hard-core scientistic project of providing an authoritative basis for the moral values that govern the behavior of man. Especially ethology, the very discipline devoted to the study of animal behavior, can be readily summoned for this purpose, as was indeed done by one of its founders, Konrad Lorenz. The ethological approach to ethics is to assign moral goodness to those righteous features of human behavior, such as altruism, mother-love, and marital fidelity, for which analogs can be found in the animal world and for whose functional role in nature credible explanations can be offered. Moreover, moral badness is assigned to those depraved features of human behavior, such as cannibalism or murder, which animals seem to avoid in the wild and exhibit only under the socio-pathological conditions of captivity. Although this procedure is used mainly for the rationalization of conventional values traditionally justified on religious grounds (as was done by Wolfgang Wickler in an—undoubtedly unintended—caricature of this approach), a not quite so trivial reversal of that procedure has surfaced recently. Here the ethological sanction of conventional morals is stood on its head, and the predicate "good" is given to some traditional "bad" features of human behavior, as was done for aggression by Desmond Morris and for homosexuality by R. P. Michael, on the grounds that animals exhibit them in nature, for functionally accountable reasons.

Another discipline called on to provide an authoritative source of moral values is evolutionary biology. The evolutionary approach to ethics, as exemplified by the writings of Waddington and Julian Huxley referred to in Nelkin's article, assigns moral goodness to features of human behavior, such as altruism, mother-love, and marital fidelity, which can be shown to promote the survival, or better yet, the further evolution of *H. sapiens*. Conversely, moral badness is assigned to those features, such as cannibalism or murder, which can be shown to affect adversely the survival or further evolution of the species. The cosmological idea underlying this approach is that evolution is progressive—i.e., that the condition of the Earth has been improving throughout geohistorical epochs, as more and more complex forms of life, and finally man himself, were making their appearance. Since, according to Darwin, natural selection has been responsible for this progressive history, it follows that "fitness," which enhances the chances of survival of the set of hereditary determinants associated with a particular behavioral feature, must be an objectively good quality. In the nineteenth century, Herbert Spencer was one of the main apostles of this particular version of hard-core scientism. Spencer thought that the concept "good" can be identified quite simply with "progress," and he thus provided moral support for the *laissez-faire* capitalist doctrine of "Social Darwinism." In line with the nearly universal contemporary rejection of that doctrine, Waddington stated in his 1960 book *The Ethical Animal* that Spencer's ethical "theories have been so completely discredited at this time that little further needs to be said about them." But then Waddington produces a mere casuistic variant of Spencer's hard-core evolutionistic ethics, namely, one that holds that although the notion of "good" cannot be simply identified with progress, a particular set of moral values can be judged to be good if it promotes "anagenesis," or evolutionary improvement.

On first sight, these biological approaches to hard-core scientism appear to fail on logical grounds. For the authority of science and the claims for the authenticity of its knowledge are themselves founded on the belief that scientific propositions are objective and value-free. In view of that belief, it would be clearly invalid to derive conclusions that predicate values from the value-free propositions of science. For instance, the project of an ethological yardstick of goodness would fail because no moral values can possibly be inferred from objective and value-free statements regarding the behavior of animals in their natural setting. Similarly, the derivation of goodness from the evolutionary "fitness" concept would fail because the primary value judgment on which the evolutionary ethic depends, namely, that evolution is progressive, cannot itself be deduced from any set of objective and value-free statements about the geohistorical record. Therefore, it would seem that hard-core scientistic ethics cannot, in fact, be based wholly on the objective propositions of science and must resort for their moral claims to unstated premises with hidden values. In the case of ethological and evolutionary ethics, the source of those unstated premises is not hard to identify. It is the Bible, to which the biological ethicists resort for this purpose more fundamentally than do the Fundamentalists: rather than taking their ethics directly from God's explicit commandments of *Exodus* 21-24, the biologists go back to the basics of *Genesis* 1-3. The idea that the natural behavior of animals provides a moral norm is clearly derived from the story of the explusion from the Garden of Eden, where before the Fall, Adam and Eve, still naked and nameless, lived like the other animals. And the idea that the course of evolution has been progressive, culminating in the appearance of *H. sapiens*, is equally clearly derived from the story of Creation, which has God making man in his own image as the crowning act of the sixth day. (In Chapter 7 I had already drawn attention to this metaphysical affinity between the cosmogonies of *Genesis* and of modern biology, and pointed out that from the Eastern perspective both of these Western accounts are simplistic ideas that pretend to explain what is obviously the unexplainable result of myriads of hidden causes, subcauses and conditions.)

On second sight, however, the derivation of value from scientific propositions may not be logically invalid after all, but for reasons that can give little comfort to the adherents of hard-core scientism. It is held by some contemporary philosophers, such as Thomas Kuhn and Paul K. Feyerabend, that the kind of impersonal and objective science on behalf of which authority is claimed is only a myth and does not, in fact, exist. Since scientists are human beings rather than disembodied spirits, since they necessarily interact with the phenomena they observe, and since they use ordinary language to communicate their results, they are really part of the problem rather than part of the solution. That is to say, scientists lack the status of observers external to the world of phenomena, a status they would have to have if scientific propositions were to be truly objective.

This is particularly evident in the case of biologists, for whom, as Ernst Mayr has pointed out, it is quite impossible in doing their work to avoid terms which do not imply functions, roles and values. For example, the ethological studies of insect societies resort to such terms as "queen," "worker," "soldier," "slave," and "caste." It would be quite unreasonable to ask ethologists to replace, for the sake of objectivity, these value-laden terms by an ostensibly neutral vocabulary—referring not to "queen" but to "type 248," or not to "caste" but to "social subset MNO." After all, it is precisely in the perception of a functional typology that any study of social behavior has its starting point: the typology both defines the phenomenon that is to be explained and already embodies part of the eventual explanation. Another example is provided by the "fitness" concept of the Darwinian selection theory. In ordinary English discourse, "fitness" connotes value, and it was that connotation which gave meaning to the slogan "survival of the fittest" in the service of nineteenth-century Social Darwinism. In their rejection of Social Darwinism, contemporary biologists point out that Spencer misunderstood the technical meaning of "fitness," which is supposed to represent a value-free algebraic parameter that measures the contribution made by hereditary determinants to the differential reproduction rate of organisms. Hence Darwinism would lead only to the neutral slogan of "survival of the survivors." However, the semantic problem posed by "fitness" is more complicated than this rough and ready dismissal of Spencer would suggest. Let us consider geology, which fathoms the physical evolution of the Earth. Geologists, like biologists, account for the history of our planet in terms of the forces of nature. But although in the course of that history geophysical features have come and gone, geological theories put forward to explain this succession of forms do not contain any concept equivalent to "fitness." There is no need for such a concept because we do not view geophysical evolution as progressive. Since the present continents are not seen as an "improvement" over the single continental mass from which they evolved, no explanation positing progress is called for. But for biologists steeped in the Western Judeo-Christian tradition, it would require self-denial to view biological evolution in any light other than that of progress. Not only does the truth of the Biblical idea of man as the crowning achievement of Creation seem self-apparent, but it is even difficult to deny that the swift, sharp-eyed hawk represents an improvement over lumbering, extinct *Archeopteryx*. Thus, if the Darwinian concept of "fitness" were really purged of all value content, it would lose its explanatory power for the deep question in want of explanation. That deep question is not "how did evolution happen?" but "what has made evolutionary progress possible?"

Now, if it is really the case that the propositions of science, and especially those of biology, are not value-free, then there is no necessary error in deriving moral values from them. Accordingly, the ethological and evolutionary ethics would not have to fail on logical grounds. But the idol of the uniquely authentic scientific knowledge that inspires the hard-core project of totalizing the scientific perspective in the first place would turn out to stand on feet of clay.

Soft-core Scientism

Since soft-core scientism does not claim to justify moral norms or values on scientific grounds, it escapes the logical dilemma of the hard-core. It is the version of scientism often held by philosophically more sophisticated scientists, who recognize that dilemma but nevertheless cannot help but believe that the scientific method, which proved of such tremendous help in allowing us to gain mastery over nature, ought to be of equal help in the management of human affairs. For instance, the physicist (and sometime molecular geneticist) Leo Szilard's science fiction tale *The Voice of the Dolphins* is obviously informed by that belief. Szilard himself was no stranger to the management of human affairs. In 1939 he had counseled Einstein to write the letter to President Roosevelt that induced the U.S. government to embark on the project for developing the atomic bomb, and in the immediate post-War years Szilard played a leading role in the efforts to bring atomic energy under civilian control. In *The Voice of the Dolphins*, which he wrote in 1961, Szilard envisages the founding in Vienna of an International Biological Research Institute. Contrary to their chartered scientific duties, the bright, young molecular biologists staffing the Vienna Institute intervene in the conduct of economic, political and military affairs, and thereby manage to save the world from the nuclear holocaust. Szilard's implication is clear: the same kind of clear thinking that cracked the genetic code will get us out of the mess that the muddled thinking of the politicians is always getting us into.

But the more restricted claim of soft-core scientism for the primacy of science as a guide to moral action also fails, if not on logical, then on empirical or practical grounds. One deficiency, as illustrated by a recent experience of mine at a conference held in Paris on "Biology and the Future of Man," is that it seems to be difficult to consider ethical issues touching on science while bearing in mind underlying moral values that have an other-than-scientific basis. At this conference an international panel of biologists held a discussion ostensibly devoted to defining the stage in embryonal development at which life can be said to begin. Strictly speaking, this topic seemed to concern a purely technical biological question. But the discussion was, in fact, addressed to the ethical problems posed by abortion, which at that very time the French parliament was considering legalizing. One of the panelists was the geneticist Jerome Lejeune, then a leader of the French "Right-to-Life" movement opposing passage of an abortion bill. Lejeune maintained that human life begins at the moment of fertilization of the ovum by the sperm, since it is at that moment that the future man acquires his genetic individuality. Hence abortion at any time thereafter is equivalent to murder and must not be sanctioned by law in a civilized state. Most of the other panelists seemed to be in favor of some kind of legalized abortion and took the view that human life really begins only at some later developmental stage, prior to which there cannot exist any moral obstacles to artificial termination of pregnancy. Some panelists thought life begins at the stage at which the heart muscles begin to beat rhythmically, others favored the stage at which electrical signals can first be detected in the brain, and yet others thought life really begins only at birth.

With regard to providing an insight into the ethical problems posed by abortion, the discussion was entirely futile, since no one troubled to consider the underlying moral issue—namely, the proscribed taking of human life. Both Lejeune and his adversaries based their arguments on biological knowledge gained from the study of animal embryos, without regard for the categorical difference between defining the beginning of the profane life of an animal and the beginning of the sacred life of a human being. But no biological discussion of the beginning of human life can proceed in an ethical context until we have answered the deep question of what makes human life sacred in the first place and have clarified the special status which we confer on fellow humans, as compared to other denizens of the living world. This lack of recognition of the true nature of the problem under discussion was particularly vexing because the panel was seated in the Great Amphitheater of the Sorbonne, facing a statue of René Descartes. After all, it was Descartes who had taken pains to point out that man is more than an automaton in human shape: he has a soul. Hence, when asking when human life begins in the ethical context of the abortion problem, the panelists—Cartesians one and all—should have been trying to focus on that moment when the embryo acquires a soul, or, in modern parlance, becomes a person. And that problem they could not have settled on genetic or physiological grounds.

A second, more serious deficiency of soft-core scientism is that it holds that the realization of moral aims is necessarily impeded by acts which are motivated by objectively false beliefs. Indeed, a more extreme version of this proposition makes the demonstrably false claim that a society is doomed if it bases its organization on scientific falsehoods. This claim is itself false because one can point to many societies of the past which operated in a successful and stable manner while making value judgements based on witchcraft, astrology, prophecy, and other practices which we now know to be scientifically unsound. The reason why objectively false beliefs can promote the realization of moral aims is that social relations are complex, multicausal phenomena and that any social aim can only be regarded as an optimization rather than a maximization of a set of values. This fact has been well known to the Chinese since the days of Confucius, and in the West has been generally recognized by cultural anthropologists ever since Bronislaw Malinowski pointed out early in this century that the function of myths and rites is to strengthen the traditions that help to maintain a social way of life. For instance, although the false belief of the Hopi Indians that they can bring about rain by dancing may have been harmful for their agriculture, the rain dance itself provided for a communal cohesion whose benefits may have outweighed the potential gains in crop yield which abandonment of that false belief might have produced.

The unwillingness to admit the possibility of deriving social benefits from holding of objectively false beliefs is at the root of the ongoing, mainly demagogic, dispute in the United States and Britain about research on the hereditary basis of intelligence, which we previously considered as a conflict between science and morals in Chapter 7. Both opposing sides in this dispute appear to accept the validity of the scientistic proposition that if there *were* a significant variation in the genetic contribution to intelligence between individuals, or between racial groups, then this factor ought to be taken into account in the organization of society. Since, as we saw, to the opponents of such research the mere consideration of the notion of hereditary determinants of intelligence, let alone taking it into account in social action, is a morally inadmissible underpinning of racist ideology, they deny outright the possibility of any connection between heredity and intelligence. Just like Christian Morgenstern's Palmström, they reason "pointedly, that which must not, cannot be." The proponents of research on hereditary determinants of intelligence, on the other hand, appear to be convinced that the failure to give due recognition to the existence of hereditary differences has pernicious social consequences and that, therefore, every effort must be made to identify the genetic basis of intelligence in a scientifically valid manner. This conclusion is not, however, rationally self-evident. For instance, let us consider Society A, which falsely believes that there is no hereditary contribution to intelligence (i.e., if that belief were really false) and utilizes its educational resources less efficiently than Society B, which "tracks" its pupils according to a scientifically validated familial or ethnic prognosis (i.e., if such a prognosis were possible). Cultural anthropologists might easily conclude under these circumstances that the losses sustained by Society A due to its falsely based educational system are more than outweighed *vis à vis* Society B by a greater communal cohesion, fostered by the (false) belief in innate human equality.

The most serious deficiency of soft-core scientism, however, derives from its overestimation of the power of science to provide an authoritative understanding of just those phenomena which are most relevant for the ethical domain. That is to say, the physical sciences whose propositions are the most solidly validated have the least bearing on the realization of moral aims, whereas the propositions of the human sciences, which have the most bearing on the realization of moral aims, are conspicuously devoid of objective validation. Biology occupies an intermediate position between these two extremes, with respect both to the validity and the moral relevance of its propositions. Although this difference between the laws of, say, physics and sociology is of course generally recognized, the deeper epistemological reasons why the physical sciences are "hard" and the human sciences "soft" are less widely appreciated.

Fractal Structures

One of these reasons is statistical and was previously discussed in Chapter 2, in terms of Benoit Mandelbrot's "second stage indeterminism." As was set forth in that earlier discussion, in doing his work the scientist has to recognize some common denominator, or structure, in an ensemble of events, and this structure is the phenomenon that is to be explained. An event that is unique, or at least that aspect of an event that makes it unique, cannot therefore be the subject of scientific investigation; an ensemble of unique events has no structure, and there is nothing in it to explain. Such events are random, and the observer perceives them as noise. Now since every real event incorporates some element of uniqueness, every ensemble of real events contains some noise. Thus the basic problem of any scientific analysis is to recognize in an ensemble of events a significant structure that stands above its inevitable background noise. Most of the phenomena for which successful scientific theories had been worked out prior to about one hundred years ago are relatively noise-free. Such phenomena were explained in terms of deterministic laws, which assert that a given initial structure can lead to one and only one final structure. But toward the end of the nineteenth century the methods of mathematical statistics came to be trained on previously inscrutable phenomena involving an appreciable element of noise. This development gave rise to the appearance of indeterministic laws of physics, such as the kinetic theory of gases and quantum mechanics. These laws of first stage indeterminism envisage that a given initial structure can lead to several alternative final structures. An indeterministic law is not devoid of predictive value, however, because to each of the several alternative final structures there is assigned a probability of its realization. Indeed, a deterministic law can be regarded as a limiting case of a more general indeterministic law in which the chance of the occurrence of one of the alternative final structures approaches certainty.

But, as was pointed out by Mandelbrot, many of those phenomena that continue to elude our successful theoretical understanding are not only inaccessible to analysis by deterministic theories, but have proven refractory also to explanation in terms of indeterministic theories. According to Mandelbrot, it is the statistical character of the noise presented by these phenomena of "second stage indeterminism" which render them scientifically opaque. Although these phenomena do not give the impression of random noise and readily evoke the perception of structures, which Mandelbrot qualified recently with the adjectival neologism *fractal*, it is very difficult to ascertain whether the structure the observer believes to have perceived is real, or whether it is merely a figment of his imagination. As Mandelbrot has pointed out, spontaneous activities leading to fractal structures predominate in the basic phenomena to which the human sciences address their analysis.

Hence it is because of the intrinsically refractory statistical character of the phenomena in want of explanation that it is possible only in exceptional cases to ascertain whether the structural elements of the propositions of the human sciences represent reality or figments of the imagination. It is for just that reason that the human sciences are "soft," and their laws generally beyond the reach of validation. This is not to suggest that the human sciences are worthless enterprises and that no attention need be paid to the insights they provide. On the contrary, we cannot do without them. But these considerations do show that the scientistic claims on behalf of an authoritative role of those sciences most relevant in guiding moral action can themselves be doubted on scientific grounds.

Structuralism

That the human sciences are, in fact, unlikely to provide the authoritative guide to the realization of moral aims envisaged by soft-core scientism has come to light with the emergence of the structuralist approach to man. As we saw in the preceding chapter, structuralism not only permits propositions about behavior that are not directly inducible from overt phenomena, but it even insists that such phenomena, or surface structures, are not by themselves explainable. According to the structuralist view, the surface structures are generated by covert deep structures, inaccessible to direct observation.

Following the early lead of Freud (who seemed unaware of the radical epistemological gulf that separates the protostructuralist method he developed from physics, and who thought that his analytical psychology would pave the way for a physics of the mind), the structuralist approach has by now become dominant in many other disciplines relevant to the soft-core scientistic project of relying on science for a guide to moral action. Despite considerable differences in approach to their subject matter, the various structuralist schools active in such disciplines as cognitive psychology, ethnology, and linguistics share a distinctive feature that sets them apart from other students of behavior. In contrast, for instance, to the ethologists mentioned earlier in this chapter, for whom the identification of the functional role of behavior has the status of an explanation, structuralists consider such functional explanations as trivial or superficial and seek to gain a more profound understanding in terms of universal deep structures. But, as was stated in the preceding chapter, the great weakness of the theoretical approach to human behavior offered by the structuralist schools is that the propositions which they offer regarding the deep structures are usually beyond the possibility of validation. The reason for this softness is not only the fractal nature of the surface structures that are perceived, which is troublesome for *every* approach to the human sciences, but also the nearly limitless flexibility of the transformational rules by which surface and deep structures are related, a difficulty with which non-structuralist approaches need not contend. That is to say, structuralist propositions about deep structures are nearly impossible to test critically by empirical study of overt surface structures, since it is almost always possible to reconcile any apparent contradiction between theory and observation by an appropriate adjustment of the transformational rules.

Science and Literature

Recognizing the fundamental softness of the structuralist theories, many adherents of soft-core scientism used to the more rigorous standards of harder scientific disciplines summarily dismiss this approach to the understanding of man. An excellent example of such a summary dismissal of a structuralist movement, namely of psychoanalysis, can be found in Medawar's *The Hope of Progress*. In a chapter entitled "Science and Literature," Medawar points out that these two important domains of creativity are not, as is often proclaimed, complementary and mutually sustaining endeavors to reach a common goal; instead, where they might be expected to cooperate, they compete for territories, and when literature arrives on the scene, it expels science. Thus, contrary to the more commonplace notion that when science arrives it expels art, the shoe can also be on the other foot. According to Medawar, the mistaken idea of the complementarity of science and literature is a legacy of nineteenth-century inductivist philosophers, such as Karl Pearson, and romantic poets, such as Matthew Arnold, who held that reason and imagination are antithetical, or at least alternative paths to truth, the former being the way of science and the latter the way of literature. But as Medawar points out, imagination is part and parcel of doing science: "All advances of scientific understanding begin with a speculative adventure, an imaginative preconception of what might be true . . . The conjecture is then exposed to criticism, to find out whether or not that imagined world is anything like the real one . . . Scientific reasoning is, therefore, . . . a dialogue between two voices, the one imaginative and the other critical." Nevertheless, although both science and literature thus depend on imagination, they differ in that literary reasoning is a monologue of the imaginative voice, unfettered by the nagging critical voice that asks whether what might be true is in fact the case. The supervision of scientific imagination by the critical voice demands that clarity be the chief criterion of good scientific exposition, whereas the freedom of the literary imagination from critical restraint gives voluptuary rhetoric free reign, often at the price of obscurity. Thus when the virus of uncritical imagination infects a territory of human inquiry to which science can lay claim, the scene becomes corrupt, a turf for mischief makers.

But are not men of letters, no less than men of science, after the truth? Yes, but, according to Medawar, there is a difference between scientific and poetic notions of the truth, between the truth of the laboratory and the truth of the salon: "When the word is used in a scientific context, *truth* means, of course, correspondence with reality." When used in a poetic context, however, truth may mean either "what *ought* to be," and thus be the revelation of an ideal, or "an alternative conception . . . which enriches our understanding of the actual by making us move and think and orientate ourselves in a domain wider than the actual," and thus be any "believable-in" statement. Since the poetic notion of truth does not ask for any empirical test, we need not be surprised that great difficulties arise when the lifestyle of the salon infiltrates the laboratory.

Medawar points to Freudian psychoanalysis as one of the glaring examples of the corruptive influence of the literary syndrome in a territory to which both literature and science can lay claim. The psychoanalyst does bring some kind of order to incoherence, and never (but never) being at a loss for an explanation, he may actually bring comfort or relief to his bewildered patient. He does so by building up a mythical structure around the patient which makes sense and is believable-in, regardless of whether or not it conforms to reality. But these procedures are "highly mischievous, not so much because they do harm or fail to do good, but because they represent a style of thought that will impede the growth of our understanding of mental illness."

That chapter is followed by "A Reply to Sir Peter Medawar" by the poet John Holloway. After engaging in some literary one-upmanship, noting that Medawar himself is not above an occasional flight into voluptuary rhetoric, insinuating that with self-professed friends like Sir Peter on its side, literature needs no enemies, and pointing out that the "literary syndrome" has nothing to do with literature at all but is evidently used by Medawar as a synonym for pseudoscience, Holloway focuses on an important defect of "Science and Literature." For in his essay Medawar fails to take account of *why* there is that difference between the notions of scientific and poetic truth. The reason is that scientific truths pertain only to that part of reality about which questions can be asked that admit clear yes-or-no answers. But literature is not, in general, concerned with such matters: poetic truths are meant to extend to those other, no less important, aspects of life which upon inquiry do not admit yes-or-no answers and about which one often cannot even speak plainly or unequivocally. Indeed, if art has any function other than entertainment, then it is to speak of the unspeakable.

Holloway's reply is followed by "A Rejoinder" in which Medawar does not deal with what, to me at least, appears to be the most important point raised by Holloway. I suspect that the unwillingness to admit, or even to consider, that there are aspects of reality which we are in desperate need of understanding but for which scientific truths cannot be established is a central, albeit unstated, premise of Medawar's philosophical outlook. It is difficult to be sure of this, however, since nowhere in his essay does Medawar explain his meaning of "reality." If reality is to take in only events, then the correspondence with reality of any statement about reality can, in principle at least, always be ascertained (for some statements in the subatomic domain, even this bare-bones criterion fails to hold, of course). But if reality comprises also the causal connections between events, then it is by no means certain that correspondence between reality and the products of imagination can always be established. This uncertainty holds particularly for events that occur in territories, such as psychology and sociology, to which, according to Medawar, both science and literature can lay claim.

Medawar's failure to concede the existence of this intrinsic limitation to our scientific quest for understanding the world is also part of the implied infrastructure of his Rhadamantine "Further Comments on Psychoanalysis" appended to "Science and Literature." Here Medawar sets forth his disdain for the mischievous psychiatric quackery spawned by Freud's epigones. Because no analyst has ever demonstrably effected a cure, Medawar feels justified in relegating the whole of psychoanalytic theory to the ashcan. But Medawar does not seem to appreciate that psychoanalysis is essentially a historical theory that endeavors to explain how the succession of mental events in a person's life eventually gave rise to his actual psyche. Hence the causal connections posited by the psychoanalyst are no more susceptible of validation than are those posited by any other kind of historian; at best, they are believable-in. It is hardly fair, therefore, for Medawar to ask, as he does here, that the validity of psychoanalytic theory be judged by its success in effecting cures. For he would surely not ask that a historian of war validate his theories by winning battles. In any case, what separates much of the present-day practice of psychoanalytic therapy from the domain of ordinary medicine is not so much that the psychiatrist cannot satisfy the rigorous criteria of proof demanded of him by Medawar that it was his treatment and not something else that cured the patient. More often than not, such proof also cannot be supplied for the cure of any particular patient effected by any other sort of physician. No, the trouble with therapeutic psychoanalysis is that since the conditions from which most patients seen by psychoanalysts nowadays seek relief have no objectively definable symptoms, it follows, *a fortiori*, that there does not exist any objective criterion of a cure. This was not so in the old days when Freud treated a hysterically blind patient; once that patient could see again, a cure had certainly been effected. But even if Medawar is right in his final judgement that psychoanalysis "is an endproduct . . . like a dinosaur or a Zeppelin; no better theory can ever be erected on its ruins," it may turn out that no better theory can be erected on any other basis. As I tried to show in the preceding chapter, it is unlikely that there *is* a mammal or airplane of psychology waiting in the wings.

Toward a Structuralist Ethics

The latter-day re-emergence of the Kantian theory of knowledge and its innate categories in the guise of the structuralist approach to man provides encouragement for developing also a Kantian structuralist ethics. The purpose of such a project is not the scientistic objective to extend the authority of science to the ethical domain, but merely to illuminate the meta-ethical question of how morals are possible at all. That is to say, although biology cannot justify moral values, it may, all the same, be able to give an account of their biological basis. This account would not, of course, consist of functional explanations of the sociobiological role that morals play in human intercourse, or of the nature of the evolutionary "fitness" which they may have conferred on *H. sapiens*. Rather, instead of Darwin's recognition of the role of natural selection in evolution, the point of departure of this project would be Kant's recognition that the peculiar obligatoriness of moral principles can be explained only by their unrestricted universality, that is to say, by their independence of any existential facts. Thus, according to Kant, it is not to promote happiness, or to serve progress, but to accede to the demand of human reason that action be in accord with universal law, that we feel obligated to obey moral principles. The origin of man's innate knowledge of this law can—it goes without saying—be consigned readily to the evolutionary history of *H. sapiens*.

Structuralist ethics would, therefore, attempt to reconcile the Kantian view of the innate, and hence subjective, source of morality, with the empirical fact that there seems to be no limit to the number of significantly different social situations for which individuals produce value judgements that appear reasonable to other men. To this end, structuralist ethics would envisage that an individual's moral judgements arise by a transformational process operating on an innate ethical deep structure. But despite their subjective source, his moral judgements are not seen as arbitrary or completely idiosyncratic by others, because the innate ethical deep structure is a universal which all humans share. This ethical deep structure would be more or less equivalent to Kant's concept of the categorical imperative, which he took to govern human action independently of any desired ends, including happiness. However, the neo-Kantian feature of the structuralist approach to morality is that it would posit a more complicated, or transformational, relation between the universal categorical imperative and particular moral judgements than the direct connection evidently envisaged by Kant.

Thus, according to this structuralist concept, the extant moral systems all share certain fundamental features, of which the very notion of moral value and the meaning of the unanalyzable, undefinable concept of "good" are the most basic, because they are all rooted in the same universal ethical deep structure. Hence the elucidation of the content of the ethical deep structure ought to be the central goal of the intellectual discipline called "ethics," which aims at accounting for morality and the normative principles that govern human action. Kant thought, of course, that he had managed to identify that content as the one fundamental categorical imperative from which all specific moral duties can be derived: "Act only according to that maxim which you can will to be universal law." It seems highly plausible that this criterion of universalizability is indeed contained in the ethical deep structures, but not in the strong form enunciated by Kant. For this all-inclusive generalization would place too severe a limitation on the creative aspect of morality and limit the variety of social situations under which rational value judgements can be produced. Thus it is clear that there are few if any moral rules which we would want to be followed without exception and for whose justifiable contravention we cannot imagine scenarios.

Here we reach what would appear to be the most significant aspect of the ethical deep structure—namely, that its open-ended creative possibilities appear to be achieved at the expense of logical consistency. That is to say, whatever may be the abstract moral content of the deep structure, its nature is such that the transformations to which it is subject give rise to a set of judgements that are not necessarily logically compatible and hence are not necessarily reconcilable rationally. Indeed, the logical dilemma of hard-core scientism—namely, that it must hold two irreconcilable beliefs, that of the authority of scientific knowledge and that of the autonomy of morals—is most plausibly attributed to that aspect of our ethical makeup. In other words, the moral dilemmas and paradoxes with which we are wrestling today are not simply the result of unenlightened or irrational human attitudes but are, instead, reflections of the fundamental inconsistency of the ethical deep structure which underlies our morality in the first place. Thus the resolution of these dilemmas, if possible at all, is not likely to be achieved merely by calling attention to their existence, as I have tried to do in the first part of this chapter, but would seem to require changing human nature. But whether such a change is possible, or even desirable, or whether we can only continue to muddle along as best we can with our paradoxical endowment appears to be the central ethical question for the future.

Bibliography

Feyerabend, Paul. *Against Method*. Atlantic Highlands, N.J.: Humanities Press, 1975.

Galpérine, C. (ed.). *Biology and the Future of Man*. Paris: The Universities of Paris, 1976.

Kuhn, T. S. *The Structure of Scientific Revolutions*. Univ. Chicago Press, 1964.

Lorenz, Konrad. *On Aggression*. New York: Harcourt, Brace & World, 1966.

Malinowski, Bronislaw. "Anthropology." In *Encyclopaedia Brittanica* (13th ed.), Vol. 29, pp. 131–140, 1926.

Mandelbrot, Benoit. *Fractals: Form, Chance, and Dimension*. San Francisco: W. H. Freeman and Company, 1977.

Mayr, Ernst. "Teleological and Teleonomic, A New Analysis." *Boston Stud. Phil. Sci.* 14, 91–117 (1974).

Medawar, Sir Peter. *The Hope of Progress.* London: Methuen, 1972.

Michael, Richard P. "Bisexuality and Ethics." In F. J. Ebling (ed.), *Biology and Ethics.* London: Academic Press, 1969.

Morgenstern, Christian. "The Impossible Fact." In *Gallows Songs and Other Poems* (translated by Max Knight). Munich: Piper Verlag, 1972.

Morris, Desmond. *The Naked Ape.* New York: McGraw-Hill, 1967.

Nelkin, Dorothy. "The Science Textbook Controversies." *Scientific American,* April 1976, pp. 33–39.

Schoeck, Helmut, and James W. Wiggins (eds.). *Scientism and Values.* Princeton, N.J.: Van Nostrand, 1960.

Spencer, Herbert. *Principles of Ethics.* London: Williams and Norgate, 1892–1893.

Szilard, Leo. *The Voice of the Dolphins.* New York: Simon and Schuster, 1961.

Waddington, C. H. *The Ethical Animal.* London: Allen and Unwin, 1960.

Wickler, Wolfgang. *Die Biologie der Zehn Gebote.* Munich: Piper Verlag, 1971.

INDEX
OF
NAMES